T0343531

AN INTRODUCTION TO
APPLIED
NUMERICAL
ANALYSIS

AN INTRODUCTION TO
APPLIED
NUMERICAL
ANALYSIS

M Ali Hooshyar
University of Texas at Dallas, USA

World Scientific

NEW JERSEY · LONDON · SINGAPORE · BEIJING · SHANGHAI · HONG KONG · TAIPEI · CHENNAI

Published by

World Scientific Publishing Co. Pte. Ltd.

5 Toh Tuck Link, Singapore 596224

USA office: 27 Warren Street, Suite 401-402, Hackensack, NJ 07601

UK office: 57 Shelton Street, Covent Garden, London WC2H 9HE

Library of Congress Control Number: 2024054330

British Library Cataloguing-in-Publication Data
A catalogue record for this book is available from the British Library.

AN INTRODUCTION TO APPLIED NUMERICAL ANALYSIS

ISBN 978-981-12-9841-7 (hardcover)
ISBN 978-981-12-9842-4 (ebook for institutions)
ISBN 978-981-12-9843-1 (ebook for individuals)

For any available supplementary material, please visit
https://www.worldscientific.com/worldscibooks/10.1142/13991#t=suppl

Desk Editors: Soundararajan Raghuraman/Rok Ting Tan

Typeset by Stallion Press
Email: enquiries@stallionpress.com

To Nahid, Dina and Parvin

Preface

This book has evolved from lecture notes for an undergraduate course in numerical analysis taught over three decades. The course is usually taken by students majoring in mathematics, computer science and physical sciences. The goal of the book is to introduce students to different available numerical procedures for finding solution of linear equations, roots of polynomial equations, interpolation and approximation, numerical differentiation and integration, solution of differential equations, error analysis, and enable students to apply commonly used numerical methods in mathematics, physical sciences, biomedical sciences, and engineering.

The aim of this book has been to present materials in an informal and user-friendly manner, while motivating study of the material being covered. To achieve this aim, ample figures and numerical tables are presented to assist the reader to better understand and follow the material under consideration. To further enhance understanding of the material covered, all figures and tables appearing in the book come with MATLAB programs that generated them. Such an approach also enhances MATLAB skills and enables one to apply and extend similar MATLAB programs to other related numerical applications.

The background needed to follow materials covered is the standard calculus sequence, a sophomore level course in differential equations and some familiarity with linear algebra and computer programming. The book also provides some MATLAB commands needed to assist students develop further MATLAB skills for various numerical applications.

The author would like to acknowledge and thank friends and colleagues, in particular Professors M. Razavy and Arthur B. Weglein for their many years of collaboration, and assistance. I am also indebted and very grateful to my wife, Nahid, for her support and encouragement in writing this book.

M. Ali Hooshyar

Contents

Preface   vii

1. Preliminaries   **1**

    1.1   Some Basic MATLAB Commands 1
    1.2   Finite Number of Digits and Computational
           Issues 3
    1.3   Loss of Significant Digits 4
    1.4   Propagation of Errors 5
    1.5   Horner's Method 6

2. Solution of Linear System of Equations   **11**

    2.1   Partial Pivoting 11
    2.2   Definition of Vector Norms 14
    2.3   Definition of Matrix Norms 14
    2.4   Condition Number 15
    2.5   Rate of Convergence 17
    2.6   Solution of $Ax = b$ via Iterations 18
    2.7   Jacobi Iteration Method 19
    2.8   Gauss–Seidel Iteration Method 21

3. Roots of Nonlinear Equations   **29**

    3.1   The Fixed Point Problem for Scalar Equations . . . 29
    3.2   Roots of Scalar Equations 33

3.2a   Bolzano bisection method 33

3.2b   Newton–Raphson method 36

3.2c   Secant method. 38

3.3   Modified Newton–Raphson Method
for Multiple Roots 42

3.4   Roots and Fixed Points of a System
of Equations 43

3.5   Newton–Raphson Method for a System
of Equations 48

3.6   Scant Method for a System of Equations 51

4. Polynomial Approximation and Interpolation   **57**

4.1   Lagrange Interpolating Polynomials 58

4.2   Chebyshev Polynomials 59

4.3   Spline Interpolation 64

4.4   Least Square Fitting of Data 68

4.5   Fourier Series and Trigonometric Polynomials . . . 78

4.6   Discrete Fourier Series 82

5. Differentiation and Integration   **89**

5.1   Numerical Differentiation 89

5.2   Richardson Extrapolation Method 90

5.3   Numerical Integration 93

5.4   Composite Quadrature 99

5.5   Adoptive Quadrature 103

5.6   Singular/Improper Integrals 105

5.7   Principal Value Integrals 107

5.8   Integration Over Infinite Intervals 109

6. Initial Value Problems   **117**

6.1   Euler's Method 119

6.2   Extensions of Euler's Method 124

6.3   Heun's Method 131

6.4   Multistep and Predictor–Corrector Methods 134

6.5   Stability for General Linear Multistep
Methods 141

6.6   Higher-Order ODE and Initial Value
Problems 147

7. Boundary Value Problems **155**

  7.1   Linear Second-Order ODE Boundary Value
         Problems 155

  7.2   Finite Difference Method for Boundary
         Value Problems 163

  7.3   Shooting Method for Nonlinear Second-Order
         ODE Boundary Value Problems 166

8. Partial Differential Equations **175**

  8.1   Finite Difference Method for Elliptic
         Equations 176

  8.2   Finite Difference Method for Hyperbolic
         Equations 183

  8.3   Finite Difference Method for Parabolic
         Equations 187

  8.4   Method of Lines for Parabolic Equations 191

  8.5   Method of Lines for Hyperbolic Equations 195

References 203

Index 207

Chapter 1

Preliminaries

Since MATLAB is used very frequently in applied sciences and numerical calculations, a brief review of some basic MATLAB commands that will be used in this text are presented in this chapter. Also, since computers use numbers in base two for calculations, the issues associated with conversion to base ten and use of finite number of digits that results in limiting accuracy of computations are reviewed and discussed in this chapter.

1.1 Some Basic MATLAB Commands

1.1a To use MATLAB click on MATLAB icon and type desired commands and save them as .m files. For example, name a created file as myfile.m.

1.1b To get help or learn about available materials in MATLAB, after the prompt >> type help help.

1.1c To get started and work with vectors in MATLAB, after the prompt >> type, for example, a=[1 2 3 4 5]; b=[6 7 8 9 10]; Note that if you don't type a semicolon after a MATLAB command, associated numerical value will appear. For example typing a=[1 2 3 4 5] will show a row vector with elements 1,2,3,4 & 5. Furthermore: c=a′ will be column vector for the row vector a.

1.1d To add above vectors a & b together, after prompt >> type c=a+b.

1

1.1e To graph b vs a, write after prompt $>>$ plot(a, b). You can label horizontal and vertical axis by the variables' names: xlabel('var1') ylabel('var2').

1.1f To print two graphs on the same graph page write i.e. plot (a,b, 'k-', a, c, 'k-.') One graph will be shown as a solid line and the other will appear as a dash dot line, receptively.

1.1g Matrix is written as: d= [1 2 3; 4 5 6; 7 8 9]. This will show a 3×3 matrix.

1.1h Transpose of d is d' and multiplication of matrix e and f is g=e*f.

1.1i To find determinant, use the command: y=det(d).

1.1j Inverse of a matrix d is found by the command h=inv(d).

1.1k Eigenvalues of d are found using the command [v,w]=eig(d), where v gives eigenvectors and w eigenvalues of d, respectively.

1.1l You can find numerical value of a particular variable after you run your program by typing its name.

1.1m A way to add several numbers, i.e. $S = a_1 + a_2 + \cdots a_N$, where a_n can be scalars, vectors or a matrices is to write:

```
S=0;
for n=1:N
S=S+a_n
end;
```

1.1n To see number of rows and column of a matrix d, type size(d)

1.1o To practice, save above equations into a MATLAB file (call it any .m file you like) that is use any name ending in .m. For example, save the followings into a file called myfile.m

```
a=[1 2 3 4 5]; b=[6 7 8 9 10];
c=a+b
plot (a,b)
xlabel('var1')
ylabel('var2')
d=[1 2 3;4 5 6; 7 8 9]; e=d';
y=det(d); x=inv(d); [v,w] =eig(x); ss= size(a);
S=0; N=5;
for n=1: N
S=S + a(n);
end;
```

After saving above as myfile.m, look at the list of MATLAB files you have and click on file myfile.m to open it and modify if needed.

1.1p To run the myfile.m type after the prompt >> myfile.

1.1q If you want to see value of any of above variables, type its name.

1.2   Finite Number of Digits and Computational Issues

By necessity in any numerical calculations using a computer, one needs to work with numbers represented by finite number of digits. For example, by default MATLAB uses 16 digits of precision in calculations. As we shall see, this limitation sometimes can result in drastic reduction in computational accuracy. To see this better let's note computers use base 2, i.e. 0 & 1 (on or off) to represent a number. However, we usually work in base 10. Thus, computers convert input numbers from base 10 to base two. This conversion could be a source of error due to the necessity to work with finite number of digits. A simple example is to find $S = \sum_{n=1}^{n=1000} 0.3 = 0.3 + 0.3 \cdots 0.3$. If a machine was working with infinite number of digits then $0.010011001\cdots$ would be the binary equivalent of 0.3 and $S = 1000 * 0.3 = 300$. However, since 0.3 has to be approximated in base 2, i.e. $0.3 \approx 0.010011001$, the found summation would not be S but $\tilde{S} = \sum_{n=1}^{n=1000} 0.010011001$. Using MATLAB it turns out the difference between S and $\tilde{S}$ will be $S - \tilde{S} \approx 5.6 * 10^{-12}$. However, if we compute $S1 = \sum_{n=1}^{n=600} 0.5$, using MATLAB, we find no error in finding the summation, and $S1 = \tilde{S}1 = 300$. The reason being that 0.5 is exactly represented in base 2 as 0.1 without the need for approximation.

Above observation leads to the definition of significant digits. More specifically let us write any number in base 10, as $\tilde{Q} = \pm P * 10^{\pm n}$, where $10 > P \geq 1$, and n being an integer. With this convention, the number of digits used to represent P are referred to as the significant digits of $\tilde{Q}$. For example, number $\tilde{Q} = 0.031$ written as $3.1 * 10^{-2}$, indicates $\tilde{Q}$ is represented by two significant digits. But if $Q=1/32=0.03125$, then $\tilde{Q} = 0.031$ would be an approximation of Q to only two significant digits.

1.2a Errors in representing a number

(i) If $\tilde{Q}$ is an approximation of Q, then absolute error in representing Q is denoted by $E_q = |Q - \tilde{Q}|$.

(ii) The relative error is denoted by $R_q = \frac{|Q-\tilde{Q}|}{|Q|}$, if $Q \neq 0$.

1.2b **Types of errors:** There are two types of errors when working with finite digits. Truncation errors, and Round-off errors. They are due to the need to limit the number of digits in computation. The procedure used depends on the computer program converting bits to digits. That is if we write $\pi = 3.14159\cdots$ as $\pi = 3.1415$, we have chopped or truncated the number. But if write π as 3.1416, we have rounded the number.

1.3 Loss of Significant Digits

Due to finite number of digits used in a computation, loss of significant digits can accrue. A good example is when subtracting two numbers, especially if they are close to each other. To see this in more detail, consider

$$\sqrt{456} - \sqrt{455} = 21.354 - 21.331 = 0.023 = 2.3 * 10^{-2}$$

It should be noted that we started with 5 digits accuracy, for each number, but the result is only accurate to two significant digits. To avoid such a drastic loss of accuracy, let us write

$$\sqrt{456} - \sqrt{455} = \frac{(\sqrt{456} - \sqrt{455})(\sqrt{456} + \sqrt{455})}{\sqrt{456} + \sqrt{455}} = \frac{456 - 455}{\sqrt{456} + \sqrt{455}}$$

$$= \frac{1}{21.354 + 21.331} = 0.023427$$

Hence, by modifying the way one calculates numerically, drastic loss of significant digits could be avoided.

A good application of the above idea is when numerically finding roots of the equation

$$ax^2 + bx + c = 0 \quad \text{which are } x_{\pm} = \frac{-b \pm \sqrt{b^2 - 4ac}}{2a}$$

Take the case when $b > 0$ and $b^2 >> 4|ac|$. If the above well-known formula is not appropriately computed, loss of significant digits will be considerable, since $x_+ = -b + b \approx 0$ and $x_- \approx -b/a$.

In order to reduce such a significant loss of digits, let us note

$$x_+ = \frac{-b + \sqrt{b^2 - 4ac}}{2a} = \frac{(-b + \sqrt{b^2 - 4ac})(-b - \sqrt{b^2 - 4ac})}{2a(-b - \sqrt{b^2 - 4ac})}$$

$$= \frac{4ac}{2a(-b - \sqrt{b^2 - 4ac})} = \frac{-2c}{b + \sqrt{b^2 - 4ac}}$$

and since we are adding b and $\sqrt{b^2 - 4ac}$ which are of similar size, loss of significant digits will not be as much. Furthermore, since $x_- + x_+ = \frac{-b}{a}$, having found x_+, enables us to find

$$x_- = \frac{-b}{a} - x_+$$

without having to face $b - \sqrt{b^2 - 4ac}$ loss of significant digits issues.

1.4 Propagation of Errors

Let us define $\epsilon_p = |p - \tilde{p}|$ and $\epsilon_q = |q - \tilde{q}|$, it then follows

$$\epsilon_{p+q} = |p + q - (\tilde{p} + \tilde{q})| \leq |p - \tilde{p}| + |q - \tilde{q}| = \epsilon_p + \epsilon_q$$

That is, $\epsilon_{p+q} \leq \epsilon_p + \epsilon_q$. In other words, for addition operations the absolute errors add.

For multiplications let us denote

$$R_{pq} = \frac{|pq - \tilde{p}\tilde{q}|}{|pq|} = \frac{|pq - \tilde{p}\tilde{q} - \tilde{p}q + \tilde{p}q|}{|pq|} = \frac{|(p - \tilde{p})q + \tilde{p}(q - \tilde{q})|}{|pq|}$$

$$\leq R_p + R_q$$

Thus for multiplication $R_{pq} \leq R_p + R_q$. In other words for multiplication, the relative errors add. It should be noted that to simplify the representation of above results, it was assumed $\frac{|\tilde{p}|}{|p|} \leq 1$.

Usually there are always some errors associated with any measurement. Thus for any computation making use of real data, the above rules for propagation of error in computation are very useful in calculating the accuracy of found result.

Definition. Symbol $O(h^n)$ stands for order of approximation. In other words:

$f(h) = p(h) + O(h^n)$ indicates existence of a finite $M > 0, \ni$ $\frac{|f(h)-p(h)|}{|h^n|} < M$. Or equivalently $|f(h) - p(h)| < M|h^n|$. This shows p is a good approximation of f if h is small and n is large.

Example. Let $g(x) = \sum_{k=0}^{n} \frac{f^{(k)}(x_0)(x-x_0)^k}{k!}$ be the n-term Taylor expansion of $f(x)$ around x_0. Hence, by Taylor theorem $f(x)-g(x) = R_{n+1} = \frac{f^{(k+1)}(\xi)(x-x_0)^{n+1}}{(n+1)!}$. Thus $|f(x) - g(x)| = O(h^{n+1})$, where $h = x - x_0$, $M = \frac{f^{(k+1)}(\xi)}{(n+1)!}$ and ξ is a number between x & x_0. Of course we don't know exact value of M since we don't know what is the value of ξ, but we know M is finite if $f^{(k+1)}$ exists.

The above example motivates study of summation methods to compute polynomials with less numerical error.

1.5   Horner's Method

The Horner's method is a procedure to find numerical sum of a polynomial with less loss of accuracy. For example consider finding

$$P_n(x) = a_n x^n + a_{n-1} x^{n-1} \cdots + a_1 x^1 + a_0$$

numerically. Since our calculations are of finite accuracy, calculating $a_n x^n$ first may become much larger (or smaller) than the other terms that need to be calculated and summed, i.e. adding $a_n x^n + a_{n-1} x^{n-1}$. With finite number of digits calculation, this summation operation may reduce accuracy of the found result. Horner's method tries to avoid this by calculating the sum in the following way

$$b_n = a_n \ \& \ b_k = b_{k+1} x + a_k, \quad \text{for } k = n - 1, n - 2, \ldots, 2, 1, 0$$

This method of adding terms in $P_n(x)$ leads to $b_0 = P_n(x)$. To see how it works take $n = 3$:

$$b_3 = a_3, \quad b_2 = b_3 x + a_2, \quad b_1 = b_2 x + a_1, \quad b_0 = b_1 x + a_0$$

Thus,

$$b_0 = (b_2 x + a_1)x + a_0 = (b_3 x + a_2)x^2 + a_1 x + a_0$$
$$= a_3 x^3 + a_2 x^2 + a_1 x + a_0 = P_3(x)$$

The advantage of Horner's method is that in each step we are adding two numbers that are not excessively different from each other.

Example. To see application of Horner's procedure, let us compute $e^{0.5}$, using Taylor expansion, by summing the result in three different ways

(i) $S_1 = \sum_{n=0}^{N}(0.5)^n/n!$
(ii) $S_2 = \sum_{n=0}^{N}(0.5)^{N-n}/(N-n)!$
(iii) $S_3 = b_0$, where $b_N = 1/N!$, $b_k = b_{k+1} * 0.5 + 1/k!$, for $k = N-1, N-2, \ldots, 2, 1, 0$.

Note in S_1 we add the terms as they appear in Taylor series. In S_2 we add Taylor series terms backwards and in S_3 we use Horner's method for finding S_3. MATLAB was then used to compute S_1, S_2, S_3 & $e^{0.5}$. The followings are numerical results for $N = 50$; $e^{0.5} = S_2 = S_3 = 1.648721270700128$ with equality being within significant digits used in MATLAB calculations. However, $e^{0.5} - S_1 = 4.4409 * 10^{-16}$, indicating a difference between $e^{0.5}$ and calculated S_1 that adds terms from left to right in the order they appear in the standard expansion formula.

This example shows the procedure used to compute can make a difference in accuracy of found numerical results. This shows that for the example considered adding the terms backwards in the sum or using Horner's method leads to a more accurate result than if terms are added in the natural order appearing in Taylor's formula.

Exercises

1.1. (a) Write a MATLAB code to compute $\sum_{n=1}^{100} 0.7$.
    (b) Is there an error in the computed result?
    (c) What is the reason if the computed result has no error?

1.2. (a) Write a MATLAB code to compute $\sum_{n=1}^{100} 0.25$.
    (b) Is there an error in the computed result?
    (c) What is the reason if the computed result has no error?

1.3. Write a MATLAB code to find solution to $A\vec{x} = \vec{b}$ when $A = [a_{ij}]$ with $a_{ii} = N$, and if $i \neq j$ then $a_{ij} = (-1)^{i+j}$, $\vec{b} = [1\ 2\ 3\ 4\ \ldots\ N]^\top$ with $i = 1, 2, 3, \ldots, N$ and $j = 1, 2, 3, \ldots, N$. Take $N = 10$.

1.4. Write a MATLAB code to graph $f(x) = e^{-x^2}$ for $-2 \leq x \leq 2$.

1.5. Let $p = \frac{1}{64}$ be approximated by $\tilde{p} = 0.0156$.

(a) To how many significant digits $\tilde{p}$ approximates p?

(b) What are the absolute and relative errors in this approximation?

1.6. Let $p = \tilde{p} + \epsilon$ and $q = \tilde{q} + \epsilon_q$, with $\tilde{p} = 2.0$, $\epsilon_p = .04$ and $\tilde{q} = 3.0$, $\epsilon_q = 0.3$. What are the absolute and relative errors of p and q? Also find R_{pq} the relative error in computing pq.

1.7. Derive the relative error $R_{\frac{p}{q}}$ formula, if $p = \tilde{p} + \epsilon_p$ and $q = \tilde{q} + \epsilon_q$.

1.8. Same as Problem 1.6, except find $R_{\frac{p}{q}}$ numerically.

1.9. Same as Problem 1.6, except numerically find $w = pq + \frac{p}{q}$ and ϵ_w.

1.10. Sometimes loss of significance error can be avoided by modifying the way a function is represented. Find an equivalent formulation of the following functions that avoids loss of significant digits.

(a) $f(x) = \sqrt{x^2 + 4} - (x + 2)$, for $x \gg 2$.

(b) $g(x) = \sin(x) - 1$, for $x \approx \frac{\pi}{2}$.

(c) $h(x) = \ln(x + 3) - \ln(x + 2)$, for $x \gg 3$.

1.11. (a) Find $x_\pm$, roots of

$$x^2 + 5000x + 1 = 0$$

first by direct evaluation of the standard equation $x_\pm = \frac{-b \pm \sqrt{b^2 - 4ac}}{2a}$.

(b) Find the solution by applying the procedure developed for the case when $b^2 \gg 4|ac|$.

(c) Compare your numerical results found by above two different procedures, parts (a.) & (b.), and find which solution satisfies $x^2 + 5000x + 1 = 0$ more accurately.

1.12. (a) Find a numerical procedure to find $x_\pm$, roots of $ax^2 + bx + c = 0$, when $b < 0$ and $b^2 \gg 4|ac|$, that reduces loss of significant digits, similar to $b > 0$ case considered in Problem 1.7.

(b) Like Problem 1.11, use your developed numerical procedure to find roots of $x^2 - 5000x + 1 = 0$ and compare these numerical solutions to solution found using the standard quadratic formula.

1.13. (a) Approximate $\frac{1}{1-x}$ using Taylor formula $f(x) = \sum_{n=0}^{N} \frac{f^{(n)}(0)}{n!} x^n$.

(b) For $N = 30$ and $x = 0.5$, find the error in this approximation.

(c) Same as (b) except for $x = 1.5$. Is the approximation satisfactory? If not why.

(d) By transforming x to $y = 1/x$, find a Taylor expansion of $\frac{1}{1-x}$, similar to part (a) that works for $x = 1.5$ and find the error for this approximation.

1.14. (a) Let $P_3(x) = a_3 x^3 + a_2 x^2 + a_1 x + a_0$ and $Q_2(x) = b_2 x^2 + b_1 x + b_0$, where $b_2 = a_3$ and $b_1 = a_2 + b_2 x_0$, $b_0 = a_1 + b_1 x_0$. Show that: $P_3(x) = Q_2(x)(x - x_0) + R$, where $R = a_0 + b_0 x_0$.

(b) Compare results in part (a) to synthetic division of $P_3(x)$ by $(x - x_0)$.

(c) Generalize the above to $P_n(x) = a_n x^n + a_{n-1} x^{n-1} + \cdots + a_1 x + a_0$. and $Q_{n-1}(x) = b_{n-1} x^{n-1} + b_{n-2} x^{n-2} + \cdots + b_1 x + b_0$, where $b_{n-1} = a_n$, $b_m = a_{m+1} + b_{m+1} x_0$, for $m = n - 2, n - 3, \ldots, 1, 0$. Show that $P_n(x) = Q_{n-1}(x)(x - x_0) + R$, where $R = a_0 + b_0 x_0$.

1.15. Use Taylor expansion to approximate $f(x) = e^x \approx \sum_{n=0}^{12} \frac{f^{(n)}(0)}{n!} x^n$ for $x = 0.8$ by adding terms in the sum in the order they appear. Then compute the same sum using Horner's method, and compare your two answers with each other and with $f(0.8)$, using MATLAB.

Chapter 2

Solution of Linear System of Equations

System of linear equations $A\mathbf{x} = \mathbf{b}$ appears in many analytical and numerical studies. There exist vast amounts of literature and numerical codes to find its solution. The most familiar approach for solving $A\mathbf{x} = \mathbf{b}$ is the Gaussian Elimination and Back substitution. However, due to the need to perform finite number of digits calculation, one needs to be careful on how to successively eliminate the unknown components of $\mathbf{x}$. A commonly used method is partial pivoting strategy to reduce rounding errors due to the need for finite number of digits computation.

2.1 Partial Pivoting

The Idea of Gaussian Elimination and back substitution is essentially to transform the matrix A into a tridiagonal form

$$
A = \begin{bmatrix} a_{11} & a_{12} & \dots & a_{1n} \\ a_{21} & a_{22} & \dots & a_{2n} \\ . & . & \dots & . \\ a_{n1} & a_{n2} & \dots & a_{nn} \end{bmatrix} \rightarrow \begin{bmatrix} 1 & \tilde{a}_{12} & \dots & \tilde{a}_{1n} \\ 0 & 1 & \dots & \tilde{a}_{2n} \\ 0 & 0 & 1.. & . \\ 0 & 0 & \dots & 1 \end{bmatrix}
$$

which is achieved by eliminating all the coefficients of the first column except for the pivot row by row operations. The process is then repeated for other columns. More specifically, the pivot row for the first column is chosen as the row with largest $|a_{k,1}|$, $k = 1, 2, \ldots, n$. That is pivot row for first column is the kth row where $|a_{k,1}| = \max\{|a_{1,1}|, |a_{2,1}|, \ldots, |a_{n-1,1}|, |a_{n,1}|\}$. Then by row operations this kth row is moved to become the first row. By appropriate constant multiplications and subtractions all other elements in column 1 are made to become zero. The procedure is similarly continued for the remaining columns successively, till the matrix become upper triangular. That is the pivot for column p is the largest $|a_{k,p}| = \max\{|a_{p,p}|, |a_{p+1,p}|, \ldots, |a_{n,p}|\}$. To see this procedure better let us consider the following linear system of equations:

$$A\mathbf{x} = \mathbf{b} = \begin{bmatrix} 10 & 1 \\ 2 & -1 \end{bmatrix} \begin{bmatrix} x_1 \\ x_2 \end{bmatrix} = \begin{bmatrix} 3 \\ 3 \end{bmatrix}$$

Since det $A = -10 - 2 \neq 0$, we know solution to above equations exists. Writing the above system of equations in its augmented form

$$\begin{bmatrix} 10 & 1 & 3 \\ 2 & -1 & 3 \end{bmatrix}, \text{ dividing its first row (R1) by } 10 \rightarrow \begin{bmatrix} 1 & 0.1 & 0.3 \\ 2 & -1 & 3 \end{bmatrix}$$

Transforming R2 to R2–2R1 (first row being the pivot and using one-digit rounding arithmetic) we find above matrix transformed to

$$\begin{bmatrix} 1 & 0.1 & 0.3 \\ 0 & -1 & 2 \end{bmatrix}. \text{ Multiplying R2 by} -1 \rightarrow \begin{bmatrix} 1 & 0.1 & 0.3 \\ 0 & 1 & -2 \end{bmatrix} \Rightarrow x_2 = -2$$

Since $x_1 + 0.1(x_2) = 0.3 \Rightarrow x_1 = 0.3 + 0.2 = 0.5$.

Let us check the answer by substituting it into our original equation

$$10x_1 + x_2 = 10x_1 - 2 = 5 - 2 = 3 \quad \text{and} \quad 2x_1 - x_2 = 1 - (-2) = 3$$

It should be noted in the above example, although we used the first row for the pivot, but in reality partial pivoting rule was followed, since $10 = |a_{1,1}| > |a_{2,1}| = 2$. To see what happens if we did not follow the pivoting rule, let us do the same example with rows reversed and take the second row as the pivot. That is write

$$10x_1 + x_2 = 3 \qquad 2x_1 - x_2 = 3$$
$$\text{as:}$$
$$2x_1 - x_2 = 3 \qquad 10x_1 + x_2 = 3$$

which is exactly the same linear system of equations we just solved. Writing the transformed equations in a matrix form leads to

$$\begin{bmatrix} 2 & -1 \\ 10 & 1 \end{bmatrix} \begin{bmatrix} x_1 \\ x_2 \end{bmatrix} = \begin{bmatrix} 3 \\ 3 \end{bmatrix} \rightarrow \begin{bmatrix} 2 & -1 & 3 \\ 10 & 2 & 3 \end{bmatrix}$$

Using first row as the pivot, dividing its first row (R1) by 2, and remembering we are using one-digit rounding arithmetic, transforms

$$\begin{bmatrix} 2 & -1 & 3 \\ 10 & 2 & 3 \end{bmatrix} \rightarrow \begin{bmatrix} 1 & -0.5 & 1 \\ 10 & 2 & 3 \end{bmatrix}$$

Next, multiply first row (R1) of above matrix by 10 and subtract it from its second row (R2) one finds

$$\begin{bmatrix} 10 & -5 & 10 \\ 10 & 2 & 3 \end{bmatrix} \rightarrow \begin{bmatrix} 1 & -0.5 & 1 \\ 0 & 7 & -7 \end{bmatrix} \rightarrow 7x_2 = -7 \rightarrow x_2 = -1$$
$$\text{and} \quad x_1 = 0.5x_2 + 1 = -0.5 + 1 = 0.5$$

Checking the answer we find

$$2x_1 - x_2 = 1 + 1 \neq 3 \quad \text{and} \quad 10x_1 + x_2 = 5 - 1 \neq 3$$

In other words the found answer, when partial pivoting rule was not followed, did not lead to the correct solution of the system of equations due to use of finite number of digits. This example also shows given a set of equations; one does not need to follow the row order as given. You can change row order according to the partial pivoting rule and successively use the rows that satisfy your partial pivoting rule.

In analytical studies, it is very common to represent solution of $Ax = b$ by $x = A^{-1}b$. It should be noted that the desired A^{-1} can also be found by Gaussian elimination using the same procedure discussed above. That is label A^{-1} as the unknown Ω, and note $AA^{-1} = \mathbb{I} \equiv A\Omega = \mathbb{I}$, where $\mathbb{I}$ is the identity matrix. Then $\Omega = A^{-1}$

is found by Gaussian elimination and back substitution. However, finding the solution to the linear system of equations $A\mathbf{x} = \mathbf{b}$, by first finding A^{-1} and then the solution $\mathbf{x} = A^{-1}\mathbf{b}$ requires more computer operations, than just finding $\mathbf{x}$ directly by using Gaussian elimination and back substitution. Thus, if one is only interested in finding the solution to a system of linear equations, it will be less time consuming if direct Gaussian elimination and back substitution is used to find the solution.

2.2 Definition of Vector Norms

The scalar symbolized by $\|.\|$ is defined to be norm of a vector if it satisfies the followings:

1. $\|\mathbf{x}\| \geq 0 \ \forall \mathbf{x}$ and $\|\mathbf{x}\| = 0$ iff $\mathbf{x} = 0$
2. $\|\alpha\mathbf{x}\| = |\alpha|\|\mathbf{x}\|$
3. $\|\mathbf{x} + \mathbf{y}\| \leq \|\mathbf{x}\| + \|\mathbf{y}\|$

Followings are the common Vector Norms used for $\mathbf{x} = (x_1, \ldots, x_n)^\top$

1. l_1 norm of $\mathbf{x}$: $l_1(\mathbf{x}) = |x_1| + \cdots + |x_n|$.
2. l_2 norm of $\mathbf{x}$: $l_2(\mathbf{x}) = (\sum_{n=1}^{n} |x_j|^2)^{1/2}$: Euclidean norm.
3. l_∞ norm of $\mathbf{x}$: $l_\infty(\mathbf{x}) = \max_j |x_j|$: Uniform norm.

2.3 Definition of Matrix Norms

For a matrix $A = [a_{ij}]$, its norm is defined as: $\|A\| = \sup_{\mathbf{x}\neq 0} \frac{\|A\mathbf{x}\|}{\|\mathbf{x}\|}$. Thus, the norm of a matrix is dependent on the chosen vector norm. For example

1. $\|A\|_1 = \max_k(\sum_{j=1}^{n} |a_{jk}|)$
2. $\|A\|_\infty = \max_j(\sum_{k=1}^{n} |a_{jk}|)$

Example. Let $A = \begin{bmatrix} 1 & -2 \\ 3 & 4 \end{bmatrix}$

1. $\|A\|_1 = \max(|a_{11}| + |a_{21}|, \ |a_{12}| + |a_{22}|) = \max(1 + 3, \ 2 + 4) = 6$
2. $\|A\|_\infty = \max(|a_{11}| + |a_{12}|, \ |a_{21}| + |a_{22}|) = \max(1 + 2, \ 3 + 4) = 7$

2.4 Condition Number

We saw if $\det A \neq 0$, then $A\mathbf{x} = \mathbf{b}$ can be solved by Gaussian elimination method if proper pivoting is used. The next computation issue to consider is what if $\det A \neq 0$ but $\det A$ is small. In this case, the problem is called ill-conditioned, which means a small change in data will make a large difference in the solution of the system of linear equations. This observation is quantified by defining the Condition number. To demonstrate the issue, consider

$$2x + 4y = 8$$
$$2x + 3.9y = 7.9$$

Writing above set of equations in a matrix form

$$A = \begin{bmatrix} 2 & 4 \\ 2 & 3.9 \end{bmatrix}, \ \det A = 2(3.9) - 2(4) = -0.2, \mathbf{b} = \begin{bmatrix} 8 \\ 7.9 \end{bmatrix}$$

One finds the exact solution to $A\mathbf{x} = \mathbf{b}$ to be: $\mathbf{x} = \begin{bmatrix} x \\ y \end{bmatrix} = \begin{bmatrix} 2 \\ 1 \end{bmatrix}$

Now, let us assume there was a small experimental error in measurement and $\mathbf{b}$ was recorded as $\tilde{\mathbf{b}} = \begin{bmatrix} 8 \\ 8 \end{bmatrix}$. In other words an error of $(8 - 7.9) * 100/7.9 \approx 1.27\%$ was made in measuring the second component of the vector $\mathbf{b}$. Thus, the measurement error is small, but how does it affect the solution to the linear equation $A\tilde{\mathbf{x}} = \tilde{\mathbf{b}}$?

Solving for $\tilde{\mathbf{x}}$ one finds

$$\tilde{\mathbf{x}} = \begin{bmatrix} \tilde{x} \\ \tilde{y} \end{bmatrix} = A^{-1}\tilde{\mathbf{b}} = \begin{bmatrix} 4 \\ 0 \end{bmatrix}$$

Comparing the result with exact $\mathbf{b}$, we find the error in the solution to be

$$\delta\mathbf{x} = \mathbf{x} - \tilde{\mathbf{x}} = \begin{bmatrix} x - \tilde{x} \\ y - \tilde{y} \end{bmatrix} = \begin{bmatrix} -2 \\ 1 \end{bmatrix}$$

Thus, the relative error for this small error in $\mathbf{b}$ will be

$$\frac{\|\delta\mathbf{x}\|_1}{\|\mathbf{x}\|_1} = \frac{|-2| + 1}{2 + 1} = 1$$

In other words, a 1.27% error in measuring the second component of $\mathbf{b}$ resulted in 100% error in the found solution.

This is one reason why most computer programs when reporting the solution $A\mathbf{x} = \mathbf{b}$, also report what is called the condition number of the matrix A, in order to provide an indication that the solution may not be reliable if A is ill conditioned.

Condition number is defined by studying how much $\mathbf{x}$ is changed if A is changed to $A + \delta A$ and $\mathbf{b} \to \mathbf{b} + \delta\mathbf{b}$. It can be shown

$$\frac{\|\delta\mathbf{x}\|}{\|\mathbf{x}\|} \le \frac{\mu}{(1 - \frac{\mu\|\delta A\|}{\|A\|})}\left(\frac{\|\delta A\|}{\|A\|} + \frac{\|\delta\mathbf{b}\|}{\|\mathbf{b}\|}\right), \text{ assuming } \|\delta A\| < \frac{1}{\|A^{-1}\|}$$

where $\mu = \mu(A) \equiv \|A\|\|A^{-1}\|$ is called the condition number and $\|A\|$ stands for the norm of matrix A. It should be noted that since

$$1 = \|I\| = \|A^{-1}A\| \le \|A^{-1}\|\|A\| = \mu(A)$$

therefore, if $\mu(A)$ is near one, the matrix A is referred to as a well-conditioned matrix and if $\mu(A)$ is exceedingly greater than one, then A is called ill-conditioned. It is apparent from the above inequalities that relative error in $\mathbf{x}$, solution of $A\mathbf{x} = \mathbf{b}$, will be in the range of relative errors of A and $\mathbf{b}$ if the condition number is small. In other words, the system is well-conditioned if condition number is small.

As an example on how to implement the above result, let us find the condition number for the matrix presented in this section using l_1 norm:

$$\|A\|_1 = \left\|\begin{bmatrix} 2 & 4 \\ 2 & 3.9 \end{bmatrix}\right\|_1 = 7.9 \ \& \ \|A^{-1}\|_1 = \left\|\begin{bmatrix} -19.5 & 20.0 \\ 10.0 & -10.0 \end{bmatrix}\right\|_1 = 30$$

Thus, $\mu(A) = \|A\|_1\|A^{-1}\|_1 = 7.98 * 30 = 239.4$. Since in the example we considered only $\mathbf{b}$ contained some measurement error, therefore $\|\delta A\|_1 = 0$ and $\|\delta\mathbf{b}\|_1 = |7.9 - 8| = 0.1$, with $\|\mathbf{b}\|_1 = 8 + 7.9 = 15.9$. Substituting these numbers in the error estimate inequality, we find:

$$\frac{\|\delta\mathbf{x}\|_1}{\|\mathbf{x}\|_1} \le \frac{\mu}{(1 - \frac{\mu\|\delta A\|_1}{\|A\|_1})}\left(\frac{\|\delta A\|_1}{\|A\|_1} + \frac{\|\delta\mathbf{b}\|_1}{\|\mathbf{b}\|_1}\right)$$

$$= 239.4\left(0 + \frac{0.1}{15.9}\right) = 1.5$$

As it can be seen from previous result, the above relative error estimate is consistent with the relative error found for the example considered, since

$$\frac{\|\delta \mathbf{x}\|_1}{\|\mathbf{x}\|_1} = 1 < 1.5$$

The above procedure was presented to demonstrate use of the inequality to estimate relative error found in the solution for problem $A\mathbf{x} = \mathbf{b}$ when there are errors in measuring A and $\mathbf{b}$. To find the estimate, we needed to find the condition number $\mu(A) = \|A\| * \|A^{-1}\|$, but if A is ill conditioned, then found A^{-1} may not be reliable. There are procedures to overcome this difficulty in estimating $\mu(A)$ without having to find A^{-1}. To become familiar with such procedures, the reader is encouraged to work through problems 2.8, 2.9 and 2.10.

Another example using MATLAB

Let's find A^{-1} when

$$A = \begin{bmatrix} 1 & (1 - 10^{-\alpha}) \\ 1 & 1 \end{bmatrix}$$

Using MATLAB command $C = \text{inv}(A)$. One finds for any constant $0 < \alpha < 16$, MATLAB computes matrix C without any warring. But for $\alpha = 16$ it gives a warning: "Matrix is close to singular or badly scaled. The result may not be accurate. $RCOND = 2.7755 * 10^{-17}$."

It should be noted that MATLAB's RCOND is the same as $1/\mu$. To verify this let's note $\mu = 1/RCOND = 1/(2.7755 * 10^{-17}) = 3.6029 * 10^{16} \approx \|A\|_\infty \|A^{-1}\|_\infty = (1 + 1)(9 + 9)10^{15} = 3.6 * 10^{16}$ which is in agreement with MATLAB value of $\mu = 1/RCOND = 3.6029 * 10^{16}$.

2.5   Rate of Convergence

Let $\{\mathbf{x}_n\}$ be a convergent sequence with $\lim_{n\to\infty} \mathbf{x}_n = \vec{\alpha}$. If there exists two non-zero positive finite constants σ and Λ such that

$$\lim_{n\to\infty} \Lambda_n(\sigma) \equiv \lim_{n\to\infty} \frac{\|\vec{\alpha} - \mathbf{x}_{n+1}\|}{\|\vec{\alpha} - \mathbf{x}_n\|^\sigma} = \Lambda$$

then σ is referred to as rate of convergence of $\{\mathbf{x}_n\}$.

Since $\vec{\alpha}$ is usually not known a priory, above defined rate of convergence is not practical for numerically estimating σ. A numerically practical way to estimate σ, is to define $\vec{\gamma}_n = (\vec{\alpha} - \mathbf{x}_n)$, and evaluate

$$\Lambda_n(\sigma) \equiv \frac{\|\mathbf{x}_{n+2} - \mathbf{x}_{n+1}\|}{\|\mathbf{x}_{n+1} - \mathbf{x}_n\|^\sigma} = \frac{\|\mathbf{x}_{n+2} - \vec{\alpha} + \vec{\alpha} - \mathbf{x}_{n+1}\|}{\|\mathbf{x}_{n+1} - \vec{\alpha} + \vec{\alpha} - \mathbf{x}_n\|^\sigma}$$

$$= \frac{\|\vec{\gamma}_{n+2} - \vec{\gamma}_{n+1}\|}{\|\vec{\gamma}_{n+1} - \vec{\gamma}_n\|^\sigma} = \frac{\|\vec{\gamma}_{n+1}\|}{\|\vec{\gamma}_n\|^\sigma} \frac{\left\| \frac{\vec{\gamma}_{n+2}}{\|\vec{\gamma}_{n+1}\|} - \frac{\vec{\gamma}_{n+1}}{\|\vec{\gamma}_{n+1}\|} \right\|}{\left\| \frac{\vec{\gamma}_{n+1}}{\|\vec{\gamma}_n\|} - \frac{\vec{\gamma}_n}{\|\vec{\gamma}_n\|} \right\|^\sigma}.$$

By definition $\|\vec{\gamma}_{n+1}\| = \Lambda \|\vec{\gamma}_n\|^\sigma$ as $n \to \infty$. Thus, $\lim_{n\to\infty} \frac{\|\vec{\gamma}_{n+1}\|}{\|\vec{\gamma}_n\|} = 0$ for $\sigma > 1$, and

$$\lim_{n\to\infty} \Lambda_n(\sigma) = \lim_{n\to\infty} \frac{\|\vec{\gamma}_{n+1}\|}{\|\vec{\gamma}_n\|^\sigma} \frac{\left\| \frac{\vec{\gamma}_{n+2}}{\|\vec{\gamma}_{n+1}\|} - \frac{\vec{\gamma}_{n+1}}{\|\vec{\gamma}_{n+1}\|} \right\|}{\left\| \frac{\vec{\gamma}_{n+1}}{\|\vec{\gamma}_n\|} - \frac{\vec{\gamma}_n}{\|\vec{\gamma}_n\|} \right\|^\sigma} = \lim_{n\to\infty} \frac{\|\vec{\gamma}_{n+1}\|}{\|\vec{\gamma}_n\|^\sigma} \frac{\frac{\|\vec{\gamma}_{n+1}\|}{\|\vec{\gamma}_{n+1}\|}}{\frac{\|\vec{\gamma}_n\|^\sigma}{\|\vec{\gamma}_n\|^\sigma}}$$

$$= \lim_{n\to\infty} \frac{\|\vec{\gamma}_{n+1}\|}{\|\vec{\gamma}_n\|^\sigma} = \lim_{n\to\infty} \frac{\|\vec{\alpha} - \mathbf{x}_{n+1}\|}{\|\vec{\alpha} - \mathbf{x}_n\|^\sigma} = \Lambda$$

In other words,

$$\lim_{n\to\infty} \frac{\|\mathbf{x}_{n+2} - \mathbf{x}_{n+1}\|}{\|\mathbf{x}_{n+1} - \mathbf{x}_n\|^\sigma} = \lim_{n\to\infty} \frac{\|\vec{\alpha} - \mathbf{x}_{n+1}\|}{\|\vec{\alpha} - \mathbf{x}_n\|^\sigma} = \Lambda.$$

The left-hand side of above equation can be numerically calculated for a selected σ. Thus, a numerical estimate of the rate of convergence can be found by choosing σ in such a way that the ratio

$$\frac{\|\mathbf{x}_{n+2} - \mathbf{x}_{n+1}\|}{\|\mathbf{x}_{n+1} - \mathbf{x}_n\|^\sigma}$$

tends to a nonzero finite positive constant. For cases when $\sigma = 1$, the sequence is said to converge linearly, and if $\sigma = 2$ the convergence is called Quadratically convergent.

2.6 Solution of $A\mathbf{x} = \mathbf{b}$ via Iterations

When solving $A\mathbf{x} = \mathbf{b}$, if the matrix A has large number of columns and rows, but most of its entries are zeros, then it is called a sparse matrix. In such cases using above mentioned procedures may not be very practical way to find the solution, since for example if $n = 10^5$, then $n^3 = 10^{15}$ operations needed to find A^{-1}. In such cases iteration methods become more practical, and work well if rows of A are strictly diagonally dominated. That is

$$|a_{kk}| > |a_{k,1}| + \cdots |a_{k,k-1}| + |a_{k,k+1}| + \cdots |a_{k,N}|, \quad k = 1, 2, 3, \ldots, N$$

Then $A\mathbf{x} = \mathbf{b}$ can be solved by iteration of $\mathbf{x}^{(k)} = (x_1^{(k)}, x_2^{(k)}, \ldots, x_N^{(k)})^\top$. Some of such iteration methods are presented in the following sections.

2.7 Jacobi Iteration Method

If A is diagonally dominated, we write $A\mathbf{x} = \mathbf{b}$ in terms of its components

$$a_{j1}x_1 + \cdots + a_{jj}x_j + \cdots a_{jN}x_N = b_j$$

then one uses this relation to find x_j in terms of other x_i iteratively. To formalize this idea in more details, let us note that the solution to this system of linear equations $\mathbf{x} = (x_1, x_2, \ldots, x_N)^\top$ needs to satisfy

$$x_j = \frac{b_j}{a_{jj}} - \sum_{i \neq j, \, i=1}^{N} \frac{a_{ji}}{a_{jj}} x_i, \quad j = 1, 2, \ldots, N$$

Note $a_{jj} \neq 0$ due to A being diagonally dominated. This rewriting of the system of linear equations lends itself easily to the following iterative solution

$$x_j^{(k+1)} = \frac{b_j}{a_{jj}} - \sum_{i \neq j, \, i=1}^{N} \frac{a_{ji}}{a_{jj}} x_i^{(k)}, \quad j = 1, 2, \ldots, N$$

called Jacobi iteration method, which only needs an initial guess $\mathbf{x}^{(0)}$ for the solution to start the iteration. Then update the solution by finding $\mathbf{x}^{(k)}$ for $k = 1, 2 \ldots$. In order to simplify the presentation let us write this iterative procedure in a vector form by defining

$$\tilde{\mathbf{b}} = \left[\tilde{b}_j\right], \quad \tilde{A} = [\tilde{a}_{ji}] \quad \text{with } \tilde{b}_j = \frac{b_j}{a_{jj}}, \quad \tilde{a}_{ji} = \frac{a_{ji}}{a_{jj}}(1 - \delta_{ji}),$$

where δ_{ji} is the Kronecker delta. The vector version of Jacobi iterative sequence takes the form

$$\mathbf{x}^{(k+1)} = \tilde{\mathbf{b}} - \tilde{A}\mathbf{x}^{(k)}$$

representing all components of the Jacobi iteration sequence defined by

$$x_j^{(k+1)} = \frac{b_j}{a_{jj}} - \sum_{i \neq j, \, i=1}^{N} \frac{a_{ji}}{a_{jj}} x_i^{(k)}, \quad j = 1, 2, \ldots, N$$

Theorem. *For diagonally dominated matrix A, the sequence* $\{\mathbf{x}^{(k)}\}$ *defined by* $\mathbf{x}^{(k+1)} = \tilde{\mathbf{b}} - \tilde{A}\mathbf{x}^{(k)}$ *converges, for any starting value of* $\mathbf{x}^{(0)}$.

Proof. Define $\mathbf{e}^{(k)} = \mathbf{x}^{(k)} - \mathbf{x}$, and apply above equations to find

$$\mathbf{x}^{(k+1)} - \mathbf{x} = \tilde{\mathbf{b}} - \tilde{\mathbf{b}} - \tilde{A}(\mathbf{x}^{(k)} - \mathbf{x})$$

$$\|\mathbf{e}^{(k+1)}\|_\infty = \|\tilde{A}\mathbf{e}^{(k)}\|_\infty \leq \|\tilde{A}\|_\infty \|\mathbf{e}^{(k)}\|_\infty$$

$$\text{where} \quad \|\tilde{A}\|_\infty = \max_j \sum_{i=1}^{N} |\tilde{a}_{ji}| = \max_j \sum_{i=1, i \neq j}^{N} \left|\frac{a_{ji}}{a_{jj}}\right| = \mu$$

and $\mu < 1$, since A is diagonally dominated. Thus,

$$\|\mathbf{e}^{(k+1)}\|_\infty \leq \mu \|\mathbf{e}^{(k)}\|_\infty \leq \mu^2 \|\mathbf{e}^{(k-1)}\|_\infty \cdots \leq \mu^{k+1} \|\mathbf{e}^{(0)}\|_\infty$$

Hence, $\lim_{k \to \infty} \mathbf{e}^{(k)} = \lim_{k \to \infty} (\mathbf{x}^{(k)} - \mathbf{x}) \leq \mu^k \|\mathbf{e}^{(0)}\|_\infty = \mathbf{0}$, implying Jacobi iteration is convergent, for any starting value $\mathbf{x}^{(0)}$, since $\mathbf{e}^{(0)} = \mathbf{x}^{(0)} - \mathbf{x}$ and $\mu < 1$. Furthermore, the smaller is μ the faster $\mathbf{x}^{(k)}$ converges to $\mathbf{x}$.

As can be noticed, in the Jacobi method, one does not make use of partially updated values of $x_j^{(k)}$ in each step of iteration. To speed up the convergence, Jacobi iteration was modified by Gauss and Seidel.

2.8 Gauss–Seidel Iteration Method

The Gauss–Seidel method is like Jacobi method, except when updating the solution, use is made of the components which are already updated. That is,

$$x_j^{(k+1)} = \frac{1}{a_{jj}} \left(b_j - \sum_{i=1}^{j-1} a_{ji} \, x_i^{(k+1)} - \sum_{i=j+1}^{N} a_{ji} \, x_i^{(k)} \right),$$

$$j = 1, 2, 3, \ldots, N$$

Theorem. A sufficient condition for Gauss–Seidel method to converge, is for the matrix A to be diagonally dominant.

Proof of Gauss–Seidel convergence is similar to what was presented for Jacobi method. That is define $\mathbf{e}^{(k)} = \mathbf{x}^{(k)} - \mathbf{x}$, where $\mathbf{x}$ is the solution of linear system being considered. The jth component of $\mathbf{e}^{(k)}$ by definition will satisfy

$$e_j^{(k+1)} = -\sum_{i=1}^{j-1} \frac{a_{ji}}{a_{jj}} \, e_i^{(k+1)} - \sum_{i=j+1}^{N} \frac{a_{ji}}{a_{jj}} \, e_i^{(k)}$$

Define

$$\sigma_j = \sum_{i=1}^{j-1} \left| \frac{a_{ji}}{a_{jj}} \right|, \quad \gamma_j = \sum_{i=j+1}^{N} \left| \frac{a_{ji}}{a_{jj}} \right|$$

with $\sigma_1 = \gamma_N = 0$. Use the definition of uniform norm to find the l component that makes $|e_l^{(k+1)}| = \|\mathbf{e}^{(k+1)}\|_\infty$. Then above equation leads to

$$|e_l^{(k+1)}| = \|\mathbf{e}^{(k+1)}\|_\infty \leq \sigma_l \, \|\mathbf{e}^{(k+1)}\|_\infty + \gamma_l \, \|\mathbf{e}^{(k)}\|_\infty$$

In other words,

$$(1 - \sigma_l)\|\mathbf{e}^{(k+1)}\|_\infty \leq \gamma_l \, \|\mathbf{e}^{(k)}\|_\infty$$

implying

$$\|\mathbf{e}^{(k+1)}\|_\infty \leq \frac{\gamma_l}{1 - \sigma_l} \, \|\mathbf{e}^{(k)}\|_\infty \leq \nu \|\mathbf{e}^{(k)}\|_\infty$$

where

$$\nu = \max_l \frac{\gamma_l}{1 - \sigma_l}$$

Let us also note that

$$\gamma_l + \sigma_l \leq \max_j (\gamma_j + \sigma_j) = \max_j \sum_{i=1, i \neq j}^{N} \left| \frac{a_{ji}}{a_{jj}} \right| = \mu < 1$$

Thus, it follows

$$\gamma_l < 1 - \sigma_l \ \& \ \nu = \max_l \frac{\gamma_l}{1 - \sigma_l} < 1$$

Next, using similar proof as presented for Jacobi method, convergence of Gauss–Seidel method also follows.

To show Gauss–Seidel converges faster than Jacobi iteration method, we need to show $\nu < \mu$. To compare μ & ν consider

$$(\gamma_l + \sigma_l) - \frac{\gamma_l}{1 - \sigma_l} = \frac{(\gamma_l + \sigma_l)(1 - \sigma_l) - \gamma_l}{1 - \sigma_l} = \frac{\sigma_l(1 - \sigma_l - \gamma_l)}{1 - \sigma_l} > 0$$

Implying

$$\gamma_l + \sigma_l > \frac{\gamma_l}{1 - \sigma_l}, \quad \forall \, l = 1, 2, \ldots, N$$

Thus, $\mu > \nu$, and in general Gauss–Siedel iteration converges faster than Jacobi iteration method for diagonally dominant matrix A.

We now present an example to see numerical performance of Jacobi and Gauss–Seidel iteration methods

Example.

$$A = \begin{pmatrix} 2 & 0 & -1 \\ 2 & 4 & 1 \\ 1 & 2 & 5 \end{pmatrix}, \quad AX = \begin{pmatrix} 3 \\ 9 \\ 13.5 \end{pmatrix}$$

Solve the above using Gaussian elimination method

$$\begin{pmatrix} 2 & 0 & -1 & 3 \\ 2 & 4 & 1 & 9 \\ 1 & 2 & 5 & 13.5 \end{pmatrix} \xrightarrow{R_1/2} \begin{pmatrix} 1 & 0 & -0.5 & 1.5 \\ 2 & 4 & 1 & 9 \\ 1 & 2 & 5 & 13.5 \end{pmatrix}$$

$$\xrightarrow[R_3-R_1]{R_2-2R_1} \begin{pmatrix} 1 & 0 & -0.5 & 1.5 \\ 0 & 4 & 2 & 6 \\ 0 & 2 & 5.5 & 12 \end{pmatrix} \xrightarrow[R_3-R_2/2]{R_2/4} \begin{pmatrix} 1 & 0 & -0.5 & 1.5 \\ 0 & 1 & 0.5 & 1.5 \\ 0 & 0 & 4.5 & 9 \end{pmatrix}$$

$$\Rightarrow x_3 = 9/4.5 = 2$$

$$x_2 = 1.5 - 0.5(2) = 0.5$$

$$x_1 = 1.5 + 0.5(2) = 2.5$$

Next, let us note:

$$A = \begin{pmatrix} 2 & 0 & -1 \\ 2 & 4 & 1 \\ 1 & 2 & 5 \end{pmatrix}$$

is a diagonally dominant matrix, since

$$|a_{11}| = 2 \geq |a_{12}| + |a_{13}| = 1,$$
$$|a_{22}| = 4 \geq |a_{21}| + |a_{23}| = 3,$$
$$|a_{33}| = 5 \geq |a_{31}| + |a_{32}| = 3.$$

Hence, Jacobi iteration converges. As an example let us start with initial values $x_0 = (1, 1, 1)^\top$

$$x_1^{(1)} = \frac{b_1 - a_{12}x_2^{(0)} - a_{13}x_3^{(0)}}{a_{11}} = \frac{3+1}{2} = 2$$

$$x_2^{(1)} = \frac{b_2 - a_{21}x_1^{(0)} - a_{23}x_3^{(0)}}{a_{22}} = \frac{9-2-1}{4} = 1.5$$

$$x_3^{(1)} = \frac{b_3 - a_{31}x_1^{(0)} - a_{32}x_2^{(0)}}{a_{33}} = \frac{13.5-1-2}{5} = 2.1$$

The iteration is then continued, till change in new iterated values are within desired tolerance.

To show how Gauss–Seidel method defines the iteration, let's note:

$$x_1^{(1)} = \frac{b_1 - a_{12}x_2^{(0)} - a_{13}x_3^{(0)}}{a_{11}} = \frac{3+1}{2} = 2$$

$$x_2^{(1)} = \frac{b_2 - a_{21}x_1^{(1)} - a_{23}x_3^{(0)}}{a_{22}} = \frac{9 - 2*2 - 1}{4} = 1$$

$$x_3^{(1)} = \frac{b_3 - a_{31}x_1^{(1)} - a_{32}x_2^{(1)}}{a_{33}} = \frac{13.5 - 1*2 - 2*1}{5} = 1.9$$

Convergence of above iterations are shown numerically in the following table, where $\|d\mathbf{x}^{(n)}\|_2$ and $\|d\mathbf{y}^{(n)}\|_2$ stand for the difference from their exact values.

Table 2.1. Jacobi and Gauss–Seidel iterations results for the $A\mathbf{x}=\mathbf{b}$ example.

#	Jacobi iteration				Gauss–Seidel			
n	$x_1^{(n)}$	$x_2^{(n)}$	$x_3^{(n)}$	$\|d\mathbf{x}^{(n)}\|_2$	$y_1^{(n)}$	$y_2^{(n)}$	$y_3^{(n)}$	$\|d\mathbf{y}^{(n)}\|_2$
0	1.0000	1.0000	1.0000	1.8708	1.0000	1.0000	1.0000	1.8708
1	2.0000	1.5000	2.1000	1.1225	2.0000	1.0000	1.9000	0.7141
2	2.5500	0.7250	1.7000	0.3783	2.4500	0.5500	1.9900	0.0714
3	2.3500	0.5500	1.9000	0.1871	2.4950	0.5050	1.9990	0.0071
4	2.4500	0.6000	2.0100	0.1122	2.4995	0.5005	1.9999	0.0007
5	2.5050	0.5225	1.9700	0.0378	2.5000	0.5000	2.0000	0.0001
6	2.4850	0.5050	1.9900	0.0187	2.5000	0.5000	2.0000	0.0000
7	2.4950	0.5100	2.0010	0.0112	2.5000	0.5000	2.0000	0.0000
8	2.5005	0.5022	1.9970	0.0038	2.5000	0.5000	2.0000	0.0000
9	2.4985	0.5005	1.9990	0.0019	2.5000	0.5000	2.0000	0.0000
10	2.4995	0.5010	2.0001	0.0011	2.5000	0.5000	2.0000	0.0000
11	2.5000	0.5002	1.9997	0.0004	2.5000	0.5000	2.0000	0.0000
12	2.4998	0.5001	1.9999	0.0002	2.5000	0.5000	2.0000	0.0000
13	2.5000	0.5001	2.0000	0.0001	2.5000	0.5000	2.0000	0.0000
14	2.5000	0.5000	2.0000	0.0000	2.5000	0.5000	2.0000	0.0000

MATLAB code for iterations in Table 2.1.

```
% inputting initial information
a11=2;a12=0;a13=-1;a21=2;a22=4;a23=1;a31=1;a32=2;a33=5;
b1=3;
b2=9;b3=13.5; x1(1)=1;x2(1)=1;x3(1)=1; N=15; y1(1)=1;y2(1)=1;
y3(1)=1;z1=2.5; z2=0.5;z3=2;
% generating Jacobi and Gauss-Seidel iterative sequences
for k=1:N
```

```
x1(k+1)=(b1-a12*x2(k)-a13*x3(k))/a11;
x2(k+1)=(b2-a21*x1(k)-a23*x3(k))/a22;
x3(k+1)=(b3-a31*x1(k)-a32*x2(k))/a33;
y1(k+1)=(b1-a12*y2(k)-a13*y3(k))/a11;
y2(k+1)=(b2-a21*y1(k+1)-a23*y3(k))/a22;
y3(k+1)=(b3-a31*y1(k+1)-a32*y2(k+1))/a33;
dx(k+1)=((x1(k+1)-z1)²+(x2(k+1)-z2)²+(x3(k+1)-z3)²)⁰·⁵;
dy(k+1)=((y1(k+1)-z1)²+(y2(k+1)-z2)²+(y3(k+1)-z3)²)⁰·⁵;
end; [x1' x2' x3' dx'y1' y2' y3' dy']
```

As can be seen from Table 2.1, Gauss–Seidel converged faster. It only took 6 iterations to converge to five significant digits, whereas Jacobi iteration took 14 iterations to converge.

Exercises

2.1. (a) Solve the following linear system of equations using Gaussian elimination and back substitution without partial pivoting, using two-digit rounding arithmetic.

$$\begin{bmatrix} 0.001 & 1 \\ 1 & 1 \end{bmatrix} \begin{bmatrix} x_1 \\ x_2 \end{bmatrix} = \begin{bmatrix} 3 \\ 2 \end{bmatrix}$$

(b) Same as in part (a) except use partial pivoting and again use two-digits rounding arithmetic.

(c) Which of the above found solutions satisfy the given linear system better?

2.2. (a) Solve the following linear system of equations using Gaussian elimination and back substitution without partial pivoting, using one-digit rounding arithmetic.

$$\begin{bmatrix} 8 & 8 & 4 \\ 4 & 6 & 1 \\ 6 & 16 & 4 \end{bmatrix} \begin{bmatrix} x_1 \\ x_2 \\ x_3 \end{bmatrix} = \begin{bmatrix} 4 \\ 5 \\ 6 \end{bmatrix}$$

(b) Does the found solution satisfy the given linear system?

2.3. (a) Same as Problem 2.2, except use partial pivoting to find the solution, with one-digit rounding arithmetic.

(b) Does the found solution satisfy the given linear system?

2.4. Let vector $\mathbf{v} = [4 \ 3 \ 5 \ -9]^\top$, find $\|\mathbf{v}\|_1$ and $\|\mathbf{v}\|_\infty$.

2.5. Let $A = \begin{bmatrix} 1 & 3 \\ 5 & -8 \end{bmatrix}$, find $\|A\|_1$ and $\|A\|_\infty$.

2.6. Let $A = \begin{bmatrix} 2 & 1 & 3 \\ 4 & 3 & -5 \\ 6 & -8 & 7 \end{bmatrix}$, find $\|A\|_1$ and $\|A\|_\infty$.

2.7. (a) Use l_∞ norm to find the condition number $\mu(A)$ for

$$A = \begin{bmatrix} 2 & 1 & 3 \\ 4 & 3 & -5 \\ 6 & -8 & 7 \end{bmatrix}$$

(b) Find the solution to $A\mathbf{x} = \mathbf{b}$, when $\mathbf{b} = [4 \ 6 \ 8]^\top$. If the given A is exactly known, but each component of $\mathbf{b}$ has some error, i.e. $\tilde{\mathbf{b}} = [3.9 \ 6.2 \ 7.3]^\top$. Estimate the expected error for your computed $\mathbf{x}$, if above defined $\tilde{\mathbf{b}}$ is used in the calculation.

2.8. Let A be an $N \times N$ matrix whose condition number is $\mu(A)$. Let Let B be an $N \times N$ matrix which is singular. Prove that

$$\frac{1}{\mu(A)} \le \frac{\|A - B\|}{\|A\|}$$

Hint: Since B is singular, make use of the fact that there exists a vector $\mathbf{x}$ with $\|\mathbf{x}\| = 1$, such that $B\mathbf{x} = 0$, and study the norm of $A^{-1}(A - B)\mathbf{x}$.

2.9. Inequality given in exercise 2.8 is used to find an estimate of $\mu(A)$ for a non-singular $N \times N$ matrix A by finding $\min[\frac{\|A-B\|}{\|A\|}]$ for all singular B matrices that are also $N \times N$. Using l_1 norm estimate $\mu(A)$ for

$$A = \begin{bmatrix} 1 & 2 & 3 & 4 \\ 0 & 5 & 6 & 7 \\ 0 & 0 & 8 & 9 \\ 0 & 0 & 0 & 10 \end{bmatrix}$$

by making use of $\min[\frac{\|A-B\|}{\|A\|}]$ relation, and not finding $\|A^{-1}\|_1$.

2.10. Let $A = \begin{bmatrix} 5 & -8 \\ 1 & 3 \end{bmatrix}$.

(a) Use

$$\mu(A) = \|A^{-1}\|_\infty \|A\|_\infty$$

to find the exact condition number for A.

(b) Use

$$\frac{1}{\mu(A)} \leq \frac{\|A - B\|_\infty}{\|A\|_\infty}$$

inequality to estimate $\mu(A)$, by selecting a suitable matrix B.

(c) Compare your estimated value of $\mu(A)$ to its exact value.

2.11. It is known that the sequence $E_n = (1 + \frac{1}{n})^n$ converges to e as n tends to infinity. By numerically evaluating

$$\frac{|E_{n+2} - E_{n+1}|}{|E_{n+1} - E_n|^p}$$

ratio, find if this sequence converges linearly or with a higher order of convergence. Clearly justify your reasons.

2.12. Write the Jacobi iteration procedure for finding solution to

$$\begin{bmatrix} 4 & 1 & 2 \\ 1 & 5 & 3 \\ 2 & 3 & -6 \end{bmatrix} \mathbf{x} = \begin{bmatrix} 8 \\ 14 \\ 2 \end{bmatrix}$$

Start with $\mathbf{x}^{(0)} = \mathbf{0}$ and perform two numerical iterations of the Jacobi method.

2.13. Write the Gauss–Seidel iteration procedure for finding solution to

$$\begin{bmatrix} 8 & 2 & -3 \\ 1 & 8 & 4 \\ -2 & 1 & 8 \end{bmatrix} \mathbf{x} = \begin{bmatrix} 2 \\ 9 \\ 14 \end{bmatrix}$$

Start with $\mathbf{x}^{(0)} = \mathbf{0}$. Perform only two numerical iterations of the Gauss–Seidel method.

2.14. Apply both Jacobi and Gauss–Seidel iteration methods for finding solution to

$$\begin{bmatrix} 3 & 2 & 0 \\ 1 & 4 & -2 \\ -2 & -2 & 5 \end{bmatrix} \mathbf{x} = \begin{bmatrix} 7 \\ 3 \\ 9 \end{bmatrix}$$

starting with $\mathbf{x}^{(0)} = \mathbf{0}$ and performing the needed number of iterations till

$$\|\mathbf{x}^{(k+1)} - \mathbf{x}^{(k)}\|_\infty \le 10^{-5}$$

for both Jacobi and Gauss–Seidel iteration methods.
Determine numerically which method converged faster.

2.15. Given the following augmented matrix:

$$\begin{bmatrix} 10 & 9 & 0 & 10.9 \\ 1 & 50 & 1 & 6.2 \\ 0 & 10 & 100 & 21 \end{bmatrix}$$

(a) Verify that $\mathbf{x} = \begin{bmatrix} 1 & 0.1 & 0.2 \end{bmatrix}^\top$ is its solution.
(b) Starting with $\mathbf{x}^{(0)} = \mathbf{0}$ and performing the needed iterations carry out five Jacobi iterations to estimate the solution $\mathbf{x}$ and compute its error.
(c) Do the same as in part (b) except use Gauss–Seidel iteration method.
(d) For this problem, which method performed better?

Chapter 3

Roots of Nonlinear Equations

Many problems of science, engineering and other disciplines can be reduced to the problem of finding solution to $f(x) = 0$, when the equality is a scalar equation, or when it is a system of nonlinear equations, $\mathbf{f(x)} = \mathbf{0}$. Thus, study of roots of nonlinear equations plays a very essential role in numerical analysis. We will first study roots of nonlinear scalar functions, and then extend the results to systems of nonlinear equations.

3.1 The Fixed Point Problem for Scalar Equations

The fixed point problem may appear in many different forms, but the methods presented to solve fixed point problems will in general be applicable to all. In order to motivate the fixed point problem, let us start with the problem of finding the roots of a scalar function $f(x) = 0$. If we add x to both sides of the equation, one finds $g(x) = f(x) + x = x$. In other words, the problem of finding roots of $f(x)$ can be changed to problem of finding an x such that $g(x) = x$. Such an x where $g(x)$ maps x back into x, itself, is called a fixed point of $g(x)$ and it is a root of $f(x) = g(x) - x$.

Theorem 3.1. *If $g(x)$ is a continuous function on a closed interval $[a, b]$ and if*

$$a \leq g(x) \leq b \quad \forall x \in [a, b]$$

then $g(x)$ has a fixed point in $[a, b]$. Furthermore, if g' exist on (a, b) and $|g'(x)| \leq \alpha < 1$ for all $x \in (a, b)$, then its fixed point is unique.

29

Proof. Let $f(x) = x - g(x)$. By assumption either $f(a) = 0$ or $f(a) < 0$, similarly $f(b) = 0$ or $f(b) > 0$. If $f(a) = 0$ this implies "a" is a fixed point of $g(x)$, likewise if $f(b) = 0$ then "b" is a fixed point of $g(x)$. Thus, we only need to consider the case when $f(a) < 0$ and $f(b) > 0$. Noting $f(x)$ is continuous, allows us to apply intermediate value theorem, which leads to existence of a number "d" such that $f(d) = 0$. That is, $f(d) = d - g(d) = 0$, or $g(d) = d$. In other words, $g(x)$ has a fixed point "d", with $a \le d \le b$.

To show uniqueness, assume we have two different fixed points, x_1 and x_2 such that $g(x_1) = x_1$, $g(x_2) = x_2$, and $x_1 \ne x_2$. Applying mean value theorem

$$\frac{g(x_1) - g(x_2)}{x_1 - x_2} = g'(c) < 1$$

If $x_1 > x_2$, the above implies

$$g(x_1) - g(x_2) < x_1 - x_2$$

However, from our assumption that x_1 and x_2 are both fixed points of $g(x)$, it follows

$$x_1 - x_2 = g(x_1) - g(x_2) < x_1 - x_2$$

which is a contradiction if $x_1 > x_2$. A similar proof for the case when $x_1 < x_2$ also leads to a similar contradiction. Thus, the assumption $x_1 \ne x_2$ is false and $g(x)$ has only a unique fixed point if $|g'(x)| \le \alpha < 1 \ \forall x \in (a, b)$.

It should be noted that the above theorem does not tell us what is the numerical value of the fixed point. It only informs us of its existence and uniqueness if stated conditions are satisfied. The standard procedure to find a fixed point for a continuous g is by iteration. That is, we start with a value x_0, find $g(x_0)$ which we call it $x_1 = g(x_0)$ and continue this process to find

$$x_{n+1} = g(x_n), \quad n = 0, 1, 2, 3 \ldots.$$

This fixed point procedure generates a sequence of points $\{x_n\}$. If limit of this sequence $\{x_n\}$ exists, i.e. $\lim_{n \to \infty} x_n = \beta$, Then β is a fixed point of g, that is $g(\beta) = \beta$. To show β is a fixed point we make use of continuity of $g(x)$ and note

$$\lim_{n \to \infty} x_{n+1} = \lim_{n \to \infty} g(x_n) = g\left(\lim_{n \to \infty} x_n\right) = g(\beta) = \beta$$

Figure 3.1 shows geometrically how the procedure is used to find the fixed point of $f(x) = x^3$, which is at $x = 0$. The figure also shows one cannot find the fixed point at $x = 1$ using this iteration procedure, even if we choose x_0 to be very close to 1.

Numerical results leading to Fig. 3.1 are shown in Table 3.1.

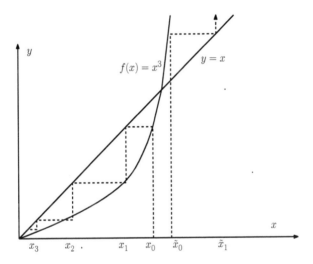

Fig. 3.1. Fixed points at $x = -1, 0, 1$.

Table 3.1. Fixed point iteration for $g(x) = x^3$, with different starting values.

n	x_{0_1}	x_{0_2}	x_{0_3}	x_{0_4}
0	0.99	1.01	−0.99	−1.01
1	0.97	1.03	−0.97	−1.03
2	0.91	1.09	−0.91	−1.09
3	0.76	1.31	−0.76	−1.31
4	0.44	2.24	−0.44	−2.24
5	0.087	11.2	−0.087	−11.2
6	0.0007	1413	−0.0007	−1413

MATLAB code for Table 3.1

```
% inputting initial information
N=6; x(1)=0.99; M=N+1;
% generating the fixed point sequence
for m=1:M
x(m+1)=[x(m)]³;
end; x'
% no semicolon after x' for printing the resulting sequence in a vector
column format.
```

Figure 3.1 and Table 3.1 both demonstrate that fixed point iteration may or may not converge. To illuminate needed conditions for a fixed point convergence, let us note that if α is a fixed point of $g(x)$ and

$$|g'(x)| \leq \gamma < 1$$

for all x in a neighborhood of α, and γ is a constant less than one, then the iteration procedure $x_{n+1} = g(x_n)$ converges to the fixed point α. To show this, mean value theorem is applied

$$|\alpha - x_1| = |g(\alpha) - g(x_0)| = |g'(\tilde{x}_0)|\,|\alpha - x_0| \leq \gamma|\alpha - x_0|$$

where x_0 is the starting value for fixed point iteration.

Applying the fixed point iteration procedure

$$|\alpha - x_n| = |g(\alpha) - g(x_{n-1})| = |g'(\tilde{x}_{n-1})|\,|\alpha - x_{n-1}|$$
$$\leq \gamma|\alpha - x_{n-1}| \leq \gamma^2|\alpha - x_{n-2}| \leq \cdots \leq \gamma^n|\alpha - x_0|$$

Above inequalities show that each newly found x_n is closer to the fixed point α, since $\gamma < 1$. Thus, if

$$|g'(x)| \leq \gamma < 1, \quad \forall x \text{ in the neighborhood of } \alpha$$

the defined fixed point iteration converges. On the other hand if

$$|g'(x)| > \beta, \quad \text{for some constant } \beta > 1$$

then each new x_n gets further away from the fixed point α and the fixed point iteration method does not converge to α, unless $x_0 = \alpha$. In that case all generated $x_n = \alpha$.

Figure 3.1 demonstrates above results graphically. Although $\alpha = 1$ is a fixed point of $f(x) = x^3$, but since $f'(x) = 3x^2$, any value for $x_o > \sqrt{1/3} \approx 0.5774$ will result in the fixed point sequence to diverge from the fixed point $\alpha = 1$, since $f'(x) > 1$. It is also interesting to note if x_0 is chosen to be 0.9, the fixed point iteration converges to the fixed point $\alpha = 0$, where in its neighborhood $|f'(x)| < 1$. Again consistent with the above findings.

Above inequality for the convergent case of fixed point iteration can also be used to estimate the number of iterations needed in finding the fixed point to the desired accuracy. To do so, let's note

$|\alpha - x_1| \leq \gamma|\alpha - x_0|$ and $\alpha - x_0 = \alpha - x_1 + x_1 - x_0$. Hence, $|\alpha - x_0| \leq |\alpha - x_1| + |x_1 - x_0| \leq \gamma|\alpha - x_0| + |x_1 - x_0| \Rightarrow |\alpha - x_0| \leq \frac{|x_1 - x_0|}{1-\gamma}$ since $0 < \gamma < 1$. Using the same procedure as used before, one finds

$$|\alpha - x_n| \leq \frac{\gamma^n |x_1 - x_0|}{1 - \gamma}$$

Since both x_0 and x_1 are known, above relation gives an estimate of error in computing α if we stop after n iteration.

Above inequality also indicates that if $\gamma < 1$ but close to 1, the iteration will converge if $|g'(x)| \leq \gamma < 1$ $\forall x$ in neighborhood of α. However, the convergence will be slow. Figure 3.1 and Table 3.1 also show the iteration procedure does not converge to fixed point $\alpha = \pm 1$, no matter how close we chose the starting point x_0 to be near fixed point $\alpha = 1$. This limitation motivates the study of more robust procedures for finding fixed points and zeros of a function.

3.2 Roots of Scalar Equations

In this section, some robust procedures to find all roots of a continuous function are presented. Some of these procedures, such as Bolzano Bisection method, make use of sectioning method of the interval of interest.

3.2a Bolzano bisection method can find all roots of a continuous function, $f(x) = 0$, and consequently all fixed points of

$g(x) = f(x) + x$. However, Bolzano bisection method has a slow rate of convergence. The method is applicable to functions that are continuous on a closed interval $a \leq x \leq b$. The method makes use of Intermediate value theorem, by noting if $\text{sign}[f(a)] \neq \text{sign}[f(b)]$, then there exists at least an $x = \beta$ such that $a < \beta < b$ and $f(\beta) = 0$. To find β one breaks the interval into two parts, by denoting its mid-point as $d = (a + b)/2$ and selecting the sub-interval $[a, d]$ or $[d, b]$, in such a way that sign of $f(x)$ at one end of the sub-interval is different from its sign at the other end. That is, if $\text{sign} [f(a)] \neq \text{sign} [f(d)]$, we choose the interval $[a, d]$ as our new interval which we call $[a_1, b_1]$. Otherwise, we choose $[d, b]$ as our new interval $[a_1, b_1]$, if $\text{sign} [f(d)] \neq \text{sign} [f(b)]$. This procedure leads to locating the root with a little more accuracy than before. This process is continued, till the root of $f(x)$, β, is determined to be in an interval as small as desired. It is evident that convergence of such a procedure is slow. Such a procedure falls in the category of Bracketing methods.

As an example let us find the fixed point of $g(x) = x^3$ located in the interval $[0.5, 2]$. Let us recall the fixed point $\alpha = 1$ could not be found by using the fixed point iteration method. Thus, let us define $f(x) = g(x) - x = x^3 - x$, and follow the Bolzano bisection

Table 3.2. Shows Bolzano bisection result for $f(x) = x^3 - x$.

n	a_n	b_n	$x_n = (b_n + a_n)/2$
0	0.5000	2.0000	1.2500
1	0.5000	1.2500	0.8750
2	0.8750	1.2500	1.0625
3	0.8750	1.0625	0.9688
4	0.9688	1.0625	1.0156
5	0.9688	1.0156	0.9922
6	0.9922	1.0156	1.0039
7	0.9922	1.0039	0.9980
8	0.9980	1.0039	1.0010
9	0.9980	1.0010	0.9995
10	0.9995	1.0010	1.0002
11	0.9995	1.0002	0.9999
12	0.9999	1.0002	1.0001
13	0.9999	1.0001	1.0000

method. Since $f(0.5) < 0$ & $f(2) > 0$, we are assured there is at least one zero in the interval $[0.5, 2]$. Next, we break the interval into two equal parts with the midpoint being at $d = (2 + 0.5)/2 = 1.25$, and test for the sign of $f(x)$ at endpoints of the sub-intervals. Noting that $f(0.5) < 0$ & $f(1.25) > 0$, but $f(1.25) > 0$ & $f(2) > 0$, leads us to focus on the interval $[0.5, 1.25]$, where at least a zero of $f(x)$ resides. The procedure is then continued, till we find an interval of sufficiently small length where the zero of $f(x)$ is located. Table 3.2 shows Bolzano bisection result for finding fixed point of $f(x) = x^3 - x$, $x \in [0.5, 2]$:

MATLAB code for Table 3.2

```
% inputting initial information
a(1)=0.5; b(1)=2;N=13; M=N+1;
% sectioning the interval, evaluating sign at ends of intervals
for k=1:N
x(k)= (b(k)+a(k))/2; sf= ((a(k))³-a(k))*((x(k))³-x(k));
if (sf < 0)
a(k+1)=a(k); b(k+1)=x(k);
else
a(k+1)=x(k); b(k+1)=b(k);
end; end;x(M)= (b(M)+a(M))/2; [a' b' x']  % printing results
```

Theorem 3.2. *Assume* $f \in C[a, b]$ *and* $f(a)$ *and* $f(b)$ *have opposite signs. Then* $\exists \ \beta \ni f(\beta) = 0$, *with* $\beta \in (a, b)$. *Furthermore, let* $\{x_n\}$ *be the sequence of mid points found for sub-intervals* $[a_n, b_n]$, $x_n = \frac{b_n + a_n}{2}$, *with* $[a_n, b_n]$ *selected in such a way that they all contain* β, *as previously described. It then follows:*

$$|\beta - x_n| \leq \frac{b - a}{2^{n+1}}, \quad n = 0, 1, 2, \ldots$$

Proof. By construction $|b_1 - a_1| = \frac{|b-a|}{2}$. Similarly for all other subintervals. Thus,

$$|b_2 - a_2| = \frac{|b_1 - a_1|}{2} = \frac{|b - a|}{2^2} \Rightarrow |b_n - a_n| = \frac{|b - a|}{2^n} \text{ and}$$

$$|\beta - x_n| \leq \frac{|b_n - a_n|}{2} = \frac{|b - a|}{2^{n+1}} \Rightarrow \lim_{n \to \infty} |\beta - x_n| = 0$$

This inequality will also provide us with an estimate of the error in finding the root if the process is stopped after N subdivisions. By construction, the error in estimating β is the length of interval $(b_N - a_N)$, containing the root β. For the example considered in Table 3.2, with $N = 13$, the error will be $L_{13} = b_{13} - a_{13} = \frac{2-0.5}{2^{14}} = 0.00009$. This estimate is consistent with results reported in Table 3.2.

Since all the methods that use bracketing procedure need to compare sign of $f(\mathrm{x})$ at end points of a line segment, thus graphing $f(x)$ is a good way to start the procedure, since it allows us to find sub-intervals that may contain a root. Then one uses a Bracketing method to more accurately isolate the numerical value of that root.

3.2b Newton–Raphson method is an iterative method for finding zeros of a function. It is intimately related to the fixed point method. In this method, one chooses a starting point x_0, finds the tangent line to $f(x)$ at x_0, and finds where tangent line intercepts the x-axis. The interception point is labeled as x_1, and x_1 is then treated as the new starting point and the process is continued. That is using the equation of tangent line $y - f(x_0) = f'(x_0)(x - x_0)$, the intercept point of the tangent line x_1 is found, and is labeled as: $x_1 = x_0 - \frac{f(x_0)}{f'(x_0)}$.

Theorem 3.3. *Let $f \in C^2[a, b]$, and $\alpha \in [a, b], \ni f(\alpha) = 0$ with $f'(\alpha) \neq 0$. Then $\exists \ \delta > 0$ such that for any $x_0 \ni |x_0 - \alpha| < \delta$, the sequence $\{x_n\}$ defined as*

$$x_n = x_{n-1} - \frac{f(x_{n-1})}{f'(x_{n-1})}$$

converges to α.

Proof. If $f(\alpha) = 0$, then

$$g(\alpha) \equiv \alpha - \frac{f(\alpha)}{f'(\alpha)} = \alpha - \frac{0}{f'(p)} = \alpha$$

Thus, the Newton–Raphson method for finding roots of $f(x)$ is equivalent to finding the fixed point of the function $g(x) = x - \frac{f(x)}{f'(x)}$.

This observation leads to convergence proof of Newton–Raphson theorem.

$$g'(x) = 1 - \frac{f'(x)}{f'(x)} + \frac{f(x)f''(x)}{f'(x)^2} = \frac{f(x)f''(x)}{f'(x)^2} \Rightarrow g'(\alpha)$$

$$= \frac{f(\alpha)f''(\alpha)}{f'(\alpha)^2} = 0$$

That is for x near a root of f, we are assured $|g'(x)| \approx 0 < 1$, and above fixed point conditions for convergence are satisfied. The assumption $f \in C^2$ implies $g'(x)$ is continuous, and for any $\gamma > 0 \; \exists \; \delta > 0 \; \ni$ if $|\alpha - x| < \delta$ then $|g'(x) - g'(\alpha)| = |g'(x)| < \gamma$. As stated before, since $|g'(x)| = \gamma < 1$, fixed point iteration results shows the sequence $\{x_n = g(x_{n-1})\}$ converges to α.

A problem with Newton–Raphson method is when $f'(x_n) \approx 0$. Then a small error in calculation can make a big difference in the sequence $\{x_n\}$, which could make the convergence slow, or problematic. Figure 3.2 shows geometrically how the method works when finding a zero of $f(x)$ using the Newton–Raphson method. It can work very well if $f'(x_n) \neq 0$, but if $f'(x_n) \approx 0$, the next itinerant

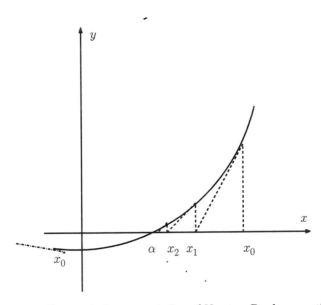

Fig. 3.2. Geometrical representation of Newton–Raphson method.

Table 3.3. Newton–Raphson results for $f(x) = x^3 - x$, with different starting values.

n	x_{0_1}	$f(x_n)$	x_{0_2}	$f(x_n)$	x_{0_3}	$f(x_n)$
0	-1.3000	-0.8970	0.4000	-0.3360	2.0000	6.0000
1	-1.0796	-0.1787	-0.2462	0.2312	1.4545	1.6228
2	-1.0080	-0.0162	0.0365	-0.0364	1.1510	0.3740
3	-1.0001	-0.0002	-0.0001	0.0001	1.0253	0.0526
4	-1.0000	-0.0000	0.0000	-0.0000	1.0009	0.0000
5	-1.0000	-0.0000	0.0000	0.0000	1.0000	0.0000

x_{n+1} may move far away from the root and produce a badly behaved, or a not converging result. Figure 3.2, also demonstrates this possible lack of convergence if we choose $x_0 \ni, f'(x_0) \approx 0$.

Table 3.3 shows Newton–Raphson method can find all the zeros of $f(x) = x^3 - x$, if starting value x_0 is close enough to the root.

MATLAB code for Table 3.3

```
% inputting initial information
N=6; x(1)=.4; f(1)=(x(1))³-x(1);
% finding Newton-Raphson sequence
for k=1:N
x(k+1)= x(k)-((x(k))³-x(k))/(3*(x(k))²-1);
f(k+1)=(x(k+1))³-x(k+1); end; [x' f']   % printing result
```

In the case when information on $f'(x)$ is not available, a natural modification of Newton–Raphson method is made by replacing $f'(x_n)$ by its finite difference $\frac{f(x_n)-f(x_{n-1})}{x_n-x_{n-1}}$. This modification of the Newton–Raphson method leads to the secant method.

3.2c   Secant method. As the name indicates, the derivative of f is replaced by its secant in the Newton–Raphson method. In other words

$$f'(x_n) \approx \frac{f(x_n) - f(x_{n-1})}{x_n - x_{n-1}}$$

Table 3.4. Solution of $f(x) = x^3 - x$ via the Secant method, with different starting values.

n	x_{01}	$f(x_n)$	x_{02}	$f(x_n)$	x_{03}	$f(x_n)$
0	−1.3000	−0.8970	0.4000	−0.3360	2.0000	6.0000
1	−1.0823	−0.1855	−0.2522	0.2362	1.4605	1.6546
2	−1.0256	−0.0531	0.0170	−0.0170	1.2550	0.7217
3	−1.0028	−0.0056	−0.0011	0.0011	1.0961	0.2207
4	−1.0001	−0.0002	0.0000	−0.0000	1.0260	0.0541
5	−1.0000	−0.0000	−0.0000	0.0000	1.0033	0.0066
6	−1.0000	−0.0000	0.0000	−0.0000	1.0001	0.0002
7	−1.0000	0	0	0	1.0000	0.0000

Substitution of this approximation in Newton–Raphson equation is called the secant method

$$x_{n+1} = x_n - \frac{f(x_n)(x_n - x_{n-1})}{f(x_n) - f(x_{n-1})}$$

The above equation indicates to compute sequence $\{x_n\}$ via the secant method, one needs to select not only starting value x_0 but also one more additional starting value x_1.

Table 3.4 shows the results for finding roots of $f(x) = x^3 - x$ via the scant method. It is instructive to compare this table with Table 3.3 that finds roots of the same function via Newtons–Raphson method.

MATLAB code for Secant method (Table 3.4)

```
% inputting initial information
N=6;  x(2)=2.5;  f(2)=(x(2))³-x(2);x(1)=x(2)*(1.01);  f(1)=(x(1))³-
x(1);
% computing the Secant sequence
for k=2:N
x(k+1)= x(k)- ((x(k))³-x(k))*(x(k)-x(h-1))/(((x(k))³-x(k-1));
f(k+1)=(x(k+1))³-x(k+1); end; [x' f']
```

Tables 3.2–3.4, all estimated roots of $f(x) = x^3 - x$. However, as it can be seen from these tables convergence rate, is slowest for Bolzano

Table 3.5. Bolzano bisection, Newton–Raphson and Secant methods rate of convergence for $f(x) = x^3 - x$, using l_2 norm.

#	Bolzano bisection		Newton–Raphson		Secant method		
n	$\Lambda_n(1)$	$\Lambda_n(2)$	$\Lambda_n(1)$	$\Lambda_n(2)$	$\Lambda_n(1)$	$\Lambda_n(1.62)$	$\Lambda_n(2)$
0	0.5000	0.6667	0.5564	1.0201	0.3808	0.5582	0.7057
1	0.5000	1.3333	0.4142	1.3649	0.7736	2.0638	3.7658
2	0.5000	2.6667	0.1942	1.5448	0.4406	1.3782	2.7723
3	0.5000	5.3333	0.0372	1.5216	0.3251	1.7355	4.6418
4	0.5000	10.6667	0.0014	1.5009	0.1388	1.6900	6.0994
5	0.5000	21.3333	0.0000	1.5001	0.0391	1.4489	12.3610

and fastest for Newton–Raphson method. It has been shown analytically that rate of convergence $\sigma = 1$ for Bolzano bisection method, $\sigma = 1.62$ for Scant method, and $\sigma = 2$ for Newton–Raphson method. These findings are tested numerically in Table 3.5 by computing

$$\Lambda_n(\sigma) = \frac{\|\mathbf{x}_{n+2} - \mathbf{x}_{n+1}\|}{\|\mathbf{x}_{n+1} - \mathbf{x}_n\|^\sigma}$$

for all the three mentioned methods.

Table 3.5 shows that:

(i) For Bolzano bisection method, $\sigma = 1$ makes $\Lambda_n(\sigma)$ to become a constant. Indicating Bolzano bisection is a linearly convergent method.

(ii) $\sigma = 2$ makes $\Lambda_n(\sigma)$ be approximately constant, which shows Newton–Raphson is a quadratically convergent method.

(iii) Table 3.5 shows neither $\sigma = 1$ nor $\sigma = 2$ make $\Lambda_n(\sigma)$ to become a constant.

However, secant iteration numbers indicate Secant method rate of convergence is $1 < \sigma < 2$. As mentioned it has been proven that $\sigma = \frac{1+\sqrt{5}}{2} \approx 1.62$ is the rate of convergence of the Secant method, consistent with results shown in Table 3.5.

Above examples also indicate that if σ is selected to be less than rate of convergence of a sequence, $\Lambda_n(\sigma)$ will tend to zero for large n values and if σ is estimated to be larger than the actual rate of convergence then $\Lambda_n(\sigma)$ will tend to infinity. Thus, by finding behavior of $\Lambda_n(\sigma)$ for different choices of σ, one is able to numerically estimate rate of convergence for a given method.

Table 3.5 example shows numerically for a simple root Newton–Raphson method converges quadratically. This motivates the question regarding rate of convergence of Newton–Raphson method for multiple roots $m > 1$. To answer this question, it turns out to be constructive if we first prove analytically that the Newton–Raphson method converges quadratically for a simple root. To prove this, let us assume all the needed conditions are satisfied so that Theorem 3.3 is applicable. To show quadratic convergence, let α be a simple zero of $f(x)$. This assumption implies $f(x)$ can be represented as

$$f(x) = (x - \alpha)q(x), \text{ with } q(\alpha) \neq 0$$

Next, apply the Newton–Raphson method

$$x_{n+1} = x_n - \frac{f(x_n)}{f'(x_n)} = x_n - \frac{(x_n - \alpha)q(x_n)}{q(x_n) + (x_n - \alpha)q'(x_n)}$$

$$x_{n+1} - \alpha = x_n - \alpha - \frac{f(x_n)}{f'(x_n)} = x_n - \alpha - \frac{(x_n - \alpha)q(x_n)}{q(x_n) + (x_n - \alpha)q'(x_n)}$$

$$\frac{x_{n+1} - \alpha}{(x_n - \alpha)^2} = \frac{1}{x_n - \alpha} - \left(\frac{1}{x_n - \alpha}\right)\frac{q(x_n)}{q(x_n) + (x_n - \alpha)q'(x_n)}$$

$$\frac{x_{n+1} - \alpha}{(x_n - \alpha)^2} = \frac{1}{x_n - \alpha} - \left(\frac{1}{x_n - \alpha}\right)\frac{1}{1 + (x_n - \alpha)\frac{q'(x_n)}{q(x_n)}}$$

$$\frac{x_{n+1} - \alpha}{(x_n - \alpha)^2} = \frac{1}{x_n - \alpha} - \left(\frac{1}{x_n - \alpha}\right)\left(1 + \sum_{k=1}^{\infty}(-1)^k(x_n - \alpha)^k\left[\frac{q'(x_n)}{q(x_n)}\right]^k\right)$$

$$\frac{x_{n+1} - \alpha}{(x_n - \alpha)^2} = \frac{q'(x_n)}{q(x_n)} - \sum_{k=2}^{\infty}(-1)^k(x_n - \alpha)^{k-1}\left[\frac{q'(x_n)}{q(x_n)}\right]^k$$

Thus,

$$\lim_{n \to \infty}\frac{|x_{n+1} - \alpha|}{|x_n - \alpha|^2} = \frac{|q'(x_n)|}{|q(x_n)|}$$

Above shows the convergence is quadratic if the starting value x_0 was close enough to α.

Similar proof can be applied when α is a zero of $f(x)$ with multiplicity of $m > 1$. In that case, $f(x)$ is represented by

$$f(x) = (x - \alpha)^m q(x), \quad \text{with} \quad q(\alpha) \neq 0.$$

Applying Newton–Raphson method to such a function leads to

$$x_{n+1} = x_n - \frac{f(x_n)}{f'(x_n)}$$

$$= x_n - \frac{(x_n - \alpha)^m q(x_n)}{m(x_n - \alpha)^{m-1} q(x_n) + (x_n - \alpha)^m q'(x_n)}$$

$$x_{n+1} - \alpha = x_n - \alpha - \frac{(x_n - \alpha) q(x_n)}{mq(x_n) + (x_n - \alpha) q'(x_n)}$$

Thus,

$$\frac{|\alpha - x_{n+1}|}{|\alpha - x_n|} = \left| 1 - \frac{q(x_n)}{mq(x_n) - (x_n - \alpha) q'(x_n)} \right|$$

Implying

$$\lim_{n \to \infty} \frac{|\alpha - x_{n+1}|}{|\alpha - x_n|} = \left(1 - \frac{1}{m} \right) \equiv \Lambda$$

By the definition of rate of convergence, it follows for multiple roots, Newton–Raphson method will converge linearly, and $|\alpha - x_{n+1}| \approx (1 - \frac{1}{m})|\alpha - x_n|$. In other words, if α is a zero of $f(x)$ with a multiplicity two, $m = 2$, each iteration reduces the error by a factor $\frac{1}{2}$, but for higher multiplicity the iteration convergence more slowly, since in each iteration the error is reduced only by a factor of $(1 - \frac{1}{m})$.

3.3 Modified Newton–Raphson Method for Multiple Roots

There are methods to restore quadratic convergence to Newton–Raphson procedure when $f(x) = 0$ has a root α with multiplicity $m > 1$. One possible approach is to define following recursion relation

$$\tilde{x}_{n+1} = \tilde{x}_n - \frac{mf(\tilde{x}_n)}{f'(\tilde{x}_n)}, \quad n = 0, 1, 2, \ldots$$

with $\{\tilde{x}_n\}$ being the sequence generated by the above modified Newton–Raphson method. Proof of this modified Newton–Raphson method converging quadratically is very similar to the proof for a simple root case. Detail of the proof is left as an exercise. The main drawback of this method is the need to know *a priori* the root's multiplicity number, before one can apply this modification. A procedure exists to estimate the multiplicity by making use of the equation

$$m = \frac{x_{n-1} - x_{n-2}}{2x_{n-1} - x_{n-2} - x_n}$$

to estimate m, as $x_n \to \alpha$. The sequence $\{x_n\}$ is found via standard Newton–Raphson procedure, with $x_0 = \tilde{x}_0$. As it can be seen the draw back of this method, is the need to first carry out the standard Newton–Raphson iteration procedure till multiplicity m is estimated, and then use the above modified Newton–Raphson method to find the quadratically convergent sequence $\{\tilde{x}_n\}$.

Another possible modification is to work with the function $g(x) = \frac{f(x)}{f'(x)}$. It follows that if α is a root of $f(x)$ with multiplicity $m > 1$, then α is a simple root of the function $g(x)$. Thus, the standard application of Newton–Raphson method to $g(x)$, leads to quadratically convergent result for α, the root of $g(x)$ and $f(x)$. From this observation, the following modified Newton–Raphson method follows:

$$x_{n+1} = x_n - \frac{f'(x_n)f(x_n)}{f'(x_n)^2 - f(x_n)f''(x_n)}$$

It is apparent that for above formula to be applicable, one needs $f(x) \in C^3$.

We have now seen several methods on how to find roots of the scalar function $f(x)$. Let us now see how these methods are extended to system of equations.

3.4   Roots and Fixed Points of a System of Equations

In many problems in science, engineering and other disciples, one finds the need to solve a system of equations, i.e. $\mathbf{F}(\mathbf{x}) = \mathbf{0}$, where $\mathbf{F}$

and $\mathbf{x}$ are vectors with N components. For example, take the case

$$F_1(x, y) = x^2 - y = 0 \quad \text{and} \quad F_2(x, y) = xy + y^2 = 0$$

which can be written as

$$\mathbf{F}(x, y) = \begin{bmatrix} F_1(x, y) \\ F_2(x, y) \end{bmatrix} = \begin{bmatrix} 0 \\ 0 \end{bmatrix} = \mathbf{0}$$

or equivalently

$$\mathbf{F}(\mathbf{x}) = \begin{bmatrix} F_1(\mathbf{x}) \\ F_2(\mathbf{x}) \end{bmatrix}$$

where $\mathbf{x} = \begin{bmatrix} x \\ y \end{bmatrix}$. Partial derivatives of the function $\mathbf{F}$ called **Jacobian** of $\mathbf{F}$ is defined by

$$J(\mathbf{F}) = \begin{bmatrix} \frac{\partial F_1}{\partial x} & \frac{\partial F_1}{\partial y} \\ \frac{\partial F_2}{\partial x} & \frac{\partial F_2}{\partial y} \end{bmatrix} = J(x, y)$$

For our example, $J(x, y) = \begin{bmatrix} 2x & -1 \\ y & (x + 2y) \end{bmatrix}$.

Like the scalar case that one denotes $df = f'(x)dx$, for a vector function $\mathbf{F}$, its variation due to change in $\mathbf{x}$ is denoted as $d\mathbf{F} = Jd\mathbf{x}$, where

$$d\mathbf{x} = \begin{bmatrix} dx \\ dy \end{bmatrix} \quad \& \quad d\mathbf{F} = \begin{bmatrix} dF_1 \\ dF_2 \end{bmatrix} = \begin{bmatrix} F_{1_x}dx + F_{1_y}dy \\ F_{2_x}dx + F_{2_y}dy \end{bmatrix} \equiv \begin{bmatrix} \Delta F_1 \\ \Delta F_2 \end{bmatrix}$$

which is a vectorized version of the scalar relation $\Delta f(x, y) = \frac{\partial f(x,y)}{\partial x}\Delta x + \frac{\partial f(x,y)}{\partial y}\Delta y$, when $\mathbf{F}$ is a column vector.

Since roots and fixed point problems for system of equations are intimately related, just like the scalar case, let us start studying the fixed point problem for vectors in two dimensions

$$\tilde{\alpha} = \mathbf{F}(\tilde{\alpha}) = \begin{bmatrix} \alpha_1 \\ \alpha_2 \end{bmatrix} = \begin{bmatrix} F_1(\alpha_1, \alpha_2) \\ F_2(\alpha_1, \alpha_2) \end{bmatrix}.$$

It is to be understood that same definition of fixed point $\vec{\alpha} = \mathbf{F}(\tilde{\alpha})$ applies for vector functions with higher number of components.

Jacobi iteration method finds the fixed point for a system of equations by selecting an initial value $\mathbf{x}^{(0)}$ and other vectors of the sequence are then generated by applying the following:

$$\mathbf{x}^{(n+1)} = \mathbf{F}(\mathbf{x}^{(n)})$$

iterative procedure. Similar to the scalar case that if $|f'| < 1$ the fixed point iteration convergences, for a system of equations the following conditions for convergence need to be satisfied.

Theorem 3.4. *If* $\mathbf{F}(\mathbf{x})$ *is continuous and its Jacobian exist for all values of* $\mathbf{x}$ *in the neighborhood of* $\vec{\alpha}$, *the fixed point of* $\mathbf{F}(\mathbf{x})$, *with the chosen* $\mathbf{x}^{(0)}$ *close enough to* $\vec{\alpha}$ *and* $\|J(\mathbf{F})\|_\infty \leq \lambda < 1$, *then* $\lim_{n \to \infty} \mathbf{x}^{(n)} = \vec{\alpha}$.

Proof of Theorem 3.4 is similar to the scalar case, if one changes absolute values appearing in the scalar case by the l_∞ norms and make use of Taylor's theorem for any two points $\mathbf{x}$ and $\mathbf{y}$ in the domain of interest leads to

$$\mathbf{f_i}(\mathbf{x}) - \mathbf{f_i}(\mathbf{y}) = \sum_{j=1}^{N} \frac{\partial f_i(\xi^{(i)})}{\partial x_j}(\mathbf{x}_j - \mathbf{y}_j), \quad i = 1, 2, 3, \ldots, N$$

where N denotes number of components, $\xi^{(i)}$ is a point on the open line segment joining $\mathbf{x}$ & $\mathbf{y}$. Since above equation is valid for each $i = 1, 2, \ldots, N$, above equation leads to

$$\|\mathbf{f}(\mathbf{x}) - \mathbf{f}(\mathbf{y})\|_\infty \leq \lambda \|\mathbf{x} - \mathbf{y}\|_\infty$$

and if $\lambda < 1$, like the proof for scalar case, Theorem 3.4 follows using similar arguments as given for the scalar case. See Isaacson and Keller (1966) for more details.

As an example on how to apply this theorem, let us take the case of a two component function with two variables. Then the Jacobi iterative sequence defined by

$$\mathbf{x}^{(n+1)} = \mathbf{F}(\mathbf{x}^{(n)}) \equiv \begin{bmatrix} x^{(n+1)} \\ y^{(n+1)} \end{bmatrix} = \begin{bmatrix} F_1(x^{(n)}, y^{(n)}) \\ F_2(x^{(n)}, y^{(n)}) \end{bmatrix}$$

converges if $\mathbf{x}^{(0)} = \begin{bmatrix} x^{(0)} \\ y^{(0)} \end{bmatrix}$ is close enough to $\vec{\alpha}$ and

$$\|J(\mathbf{F})\|_\infty = \text{Max} \left\{ \left(\left| \frac{\partial F_1}{\partial x} \right| + \left| \frac{\partial F_1}{\partial y} \right| \right), \left(\left| \frac{\partial F_2}{\partial x} \right| + \left| \frac{\partial F_2}{\partial y} \right| \right) \right\} < 1$$

for all values of $\mathbf{x}$ in the $\vec{\alpha}$ neighborhood. In other words similar to the scalar case, the Jacobi sequence $\{\mathbf{x}^{(n)}\}$ converges if the above condition is satisfied.

Gauss–Seidel method for a system of equations, is like the procedure developed in Section 2.8. It modifies the Jacobi fixed point method for a system of equations by making use of already available updated values of $\mathbf{x}$ components. For example, in a two-variable case, Gauss–Seidel modified fixed point method takes the following form:

$$\begin{bmatrix} x^{(n+1)} \\ y^{(n+1)} \end{bmatrix} = \begin{bmatrix} F_1(x^{(n)}, y^{(n)}) \\ F_2(x^{(n+1)}, y^{(n)}) \end{bmatrix}$$

Comparing the above with the Jacobi method, we note equation

$x^{(n+1)} = F_1(x^{(n)}, y^{(n)})$ is the same in both methods, but in Jacobi's method
$y^{(n+1)} = F_2(x^{(n)}, y^{(n)})$. However, in Gauss–Seidel method updates $x^{(n)}$ is used to find $y^{(n+1)} = F_2(x^{(n+1)}, y^{(n)})$.

An example for above procedures is given for the fixed point problem

$$\mathbf{x} = \mathbf{F}(\mathbf{x}) = \begin{bmatrix} (-6x^2 + 12x + 2y^2 + 1)/12 \\ (3x - 2x^2 + 6y - 2y^2 + 3)/6 \end{bmatrix} = \begin{bmatrix} x \\ y \end{bmatrix}$$

Table 3.6. Comparing fixed point & Gauss–Siedel fixed point procedures.

#	Jacobi fixed point procedure		Gauss–Siedel fixed point procedure	
n	$x(n)$	$y(n)$	$x(n)$	$y(n)$
0	3.0000	3.0000	3.0000	3.0000
1	0.0833	−1.0000	0.0833	0.5394
2	0.3299	−0.7940	0.2117	1.0333
3	0.4639	−0.3755	0.4506	1.3350
4	0.4631	0.2378	0.7294	1.4283
5	0.4486	0.8790	0.8867	1.4296
6	0.5601	1.2787	0.9175	1.4265
7	0.7591	1.4091	0.9191	1.4262
8	0.8853	1.4347	0.9191	1.4262
9	0.9198	1.4300	0.9190	1.4262
10	0.9209	1.4263	0.9190	1.4262
11	0.9192	1.4259	0.9190	1.4262
12	0.9190	1.4261	0.9190	1.4262
13	0.9190	1.4262	0.9190	1.4262

MATLAB codes for Table 3.6 fixed point example

```
% inputting initial data
xs(1)=3; ys(1)=3;xr(1)=xs(1); yr(1)=ys(1);K=13;
% finding Jacobi & Seidel sequences
for k=1:K
```
$$xr(k+1) = (-6 * (xr(k))^2 + 12 * xr(k) + 2 * (yr(k))^2 + 1)/12;$$
$$yr(k+1) = (3 * xr(k) - 2 * (xr(k))^2 + 6 * yr(k) - 2 * (yr(k))^2 + 3)/6;$$
$$xs(k+1) = (-6 * (xs(k))^2 + 12 * xs(k) + 2 * (ys(k))^2 + 1)/12;$$
$$ys(k+1) = (3 * xs(k+1) - 2 * (xs(k+1))^2 + 6 * ys(k) - 2 * (ys(k))^2 + 3)/6;\ \text{end};\ [\text{xr' yr' xs' ys'}\,]$$

As expected, Table 3.6 shows that Gauss–Seidel fixed point procedure converges faster than the standard Jacobi procedure. Also if one tries to have a starting point too far from the fixed point, for example, $x^{(0)} = y^{(0)} = 5$, then the above presented fixed point MATLAB code does not produce a convergent result.

3.5 Newton–Raphson method for a system of equations

In order to find the roots of a system of equations, Newton–Raphson method developed for the scalar case

$$x^{(k+1)} = x^{(k)} - \frac{f(x^{(k)})}{f'(x^{(k)})} = x^{(k)} - [f'(x^{(k)})]^{-1} f(x^{(k)})$$

can be extended for roots of a system of equations, by replacing $f'(x)$ with its Jacobian J of $\mathbf{F} = [F_1 \ F_2 \ \ldots \ F_N]^\top$. These modifications leads to the Newton–Raphson procedure for a system of equations:

$$\mathbf{x}^{(k+1)} = \mathbf{x}^{(k)} - \left[J(\mathbf{F}(\mathbf{x}^{(k)}))\right]^{-1} \mathbf{F}(\mathbf{x}^{(k)})$$

with elements of Jacobian of $\mathbf{F}(\mathbf{x^{(k)}})$ being:

$$J(\mathbf{F}(\mathbf{x}^{(k)})) = \left[J_{ij}(\mathbf{F}(\mathbf{x}^{(k)}))\right]$$
$$= \left[\frac{\partial F_i(x_1^{(k)}, x_2^{(k)}, \ldots, x_j^{(k)}, \ldots, x_N^{(k)})}{\partial x_j^{(k)}}\right], \quad i \ \& \ j = 1, 2, \ldots, N$$

Convergence of Newton–Raphson iteration for a system of equations follows in a similar manner as shown for the scalar case.

Theorem 3.5. *If* $\mathbf{F}(\mathbf{x}) \in C^2(D)$, $D \subset R^N$ *&* $\mathbf{F}(\vec{\alpha}) = 0$ *with* $\vec{\alpha} \in D$, *and* $\det[J(\mathbf{F}(\vec{\alpha}))] \neq 0$, *then the sequence* $\{\mathbf{x}^{(k)}\}$ *defined by* $\mathbf{x}^{(k+1)} = \mathbf{x}^{(k)} - J^{-1}(\mathbf{F}(\mathbf{x}^{(k)}))\mathbf{F}(\mathbf{x}^{(k)})$ *converges to* $\vec{\alpha}$, *if:*

(i) $J(\mathbf{x}^{(0)})$ *has a bounded inverse* $\|J^{-1}(\mathbf{x}^{(0)})\|_\infty \leq \beta$.

(ii) *Difference between the first two Newton iterates is bounded*

$$\|\mathbf{x}^{(1)} - \mathbf{x}^{(0)}\|_\infty = \|J^{-1}(\mathbf{F}(\mathbf{x}^{(0)}))\mathbf{F}(\mathbf{x}^{(0)})\|_\infty \leq \delta$$

(iii) *All components of* $\mathbf{F}$ *have continuous derivatives satisfying:*

$$\sum_{n=1}^{N} \left|\frac{\partial^2 F_i(\mathbf{x})}{\partial x_n \partial x_j}\right| \leq \frac{\lambda}{N} , \forall \ \left\|\mathbf{x} - \mathbf{x}^{(0)}\right\|_\infty \leq 2\delta, \quad i, j = 1, 2, 3, \ldots, N$$

(iv) *and if* $\beta * \lambda * \delta \leq \frac{1}{2}$, *then the Newton iterates* $\mathbf{x}^{(k)}$ *converge to* $\vec{\alpha}$ *for all* $\mathbf{x}^{(0)}$ *satisfying the inequality:* $\|\mathbf{x}^{(0)} - \vec{\alpha}\|_\infty \leq 2\delta$.

Proof. The details of proof is beyond the scope of this book, but idea for the proof has been already presented in Theorem 3.3. Readers interested in more details are referred to the text book by Isaacson and Keller. The basic idea for the proof is to define

$$\mathbf{P}(\mathbf{x}) = \mathbf{x} - J^{-1}(\mathbf{F}(\mathbf{x}))\mathbf{F}(\mathbf{x})$$

and note that $\mathbf{P}(\tilde{\alpha}) = \tilde{\alpha}$, if $\mathbf{F}(\tilde{\alpha}) = 0$. In other words $\tilde{\alpha}$, a root of $\mathbf{F}(\mathbf{x})$ is a fixed point of the function $\mathbf{P}(\mathbf{x})$. Next take derivative of $\mathbf{P}(\mathbf{x})$ respect to x_i

$$\frac{\partial \mathbf{P}(\mathbf{x})}{\partial x_i} = \frac{\partial \mathbf{x}}{\partial x_i} - \frac{\partial J^{-1}(\mathbf{F}(\mathbf{x}))}{\partial x_i}\mathbf{F}(\mathbf{x}) - J^{-1}(\mathbf{F}(\mathbf{x}))\frac{\partial \mathbf{F}(\mathbf{x})}{\partial x_i},$$
$$i = 1, 2, 3, \ldots, N$$

Implying

$$J(\mathbf{P}(\mathbf{x})) = I - J(J^{-1}(\mathbf{F}(\mathbf{x}))\mathbf{F}(\mathbf{x}) - J^{-1}(\mathbf{F}(\mathbf{x}))J(\mathbf{F}(\mathbf{x}))$$
$$= -J(J^{-1}(\mathbf{F}(\mathbf{x}))\mathbf{F}(\mathbf{x})$$

In other words

$$\frac{\partial \mathbf{P}(\mathbf{x})}{\partial x_i} = -\frac{\partial J^{-1}(\mathbf{F}(\mathbf{x}))}{\partial x_i}\mathbf{F}(\mathbf{x})$$
$$= \left[J^{-1}(\mathbf{F}(\mathbf{x}))\frac{\partial J(\mathbf{F}(\mathbf{x}))}{\partial x_i}J^{-1}(\mathbf{F}(\mathbf{x})) \right] \mathbf{F}(\mathbf{x}))$$

To arrive at the above, use the identity $J^{-1}(\mathbf{F}(\mathbf{x})))J(\mathbf{F}(\mathbf{x}))) = I$, where I is the identity matrix. Next take its derivative respect to x_i

$$\frac{\partial J^{-1}(\mathbf{F}(\mathbf{x}))J(\mathbf{F}(\mathbf{x}))}{\partial x_i} = \frac{\partial J^{-1}(\mathbf{F}(\mathbf{x}))}{\partial x_i}J(\mathbf{F}(\mathbf{x}))$$
$$+ J^{-1}(\mathbf{F}(\mathbf{x}))\frac{\partial J(\mathbf{F}(\mathbf{x}))}{\partial x_i} = 0$$

Implying

$$\frac{\partial J^{-1}(\mathbf{F}(\mathbf{x}))}{\partial x_i} = -J^{-1}(\mathbf{F}(\mathbf{x}))\frac{\partial J(\mathbf{F}(\mathbf{x}))}{\partial x_i}J^{-1}(\mathbf{F}(\mathbf{x}))$$

and the desired relation follows.

The rest is like the proof given for Theorem 3.3. That is since $\mathbf{F}(\vec{\alpha}) = 0$, and $\mathbf{F}(\mathbf{x}) \in C^2(D)$, all the second derivatives $\frac{\partial^2 \mathbf{F}(\mathbf{x})}{\partial x_i \partial x_j}$ exist and are continuous. Hence $\frac{\partial \mathbf{P}(\mathbf{x})}{\partial x_i}$ is continuous in the $\vec{\alpha}$ neighborhood. Furthermore exists a δ such that $\|\mathbf{J}(\mathbf{P}(\mathbf{x})\|_\infty \le \lambda < 1$, for $\|\mathbf{x}^{(0)} - \vec{\alpha}\|_\infty \le 2\delta$, implying convergence of Newton–Raphson iterates.

Example. To show how Newton–Raphson method is implemented, let us take the function $\mathbf{F}(\mathbf{x}) = \begin{bmatrix} (-6x^2 + 2y^2 + 1)/12 \\ (3x - 2x^2 - 2y^2 + 3)/6 \end{bmatrix}$ and find its Jacobian

$$J(\mathbf{F}(x,y)) = \begin{bmatrix} F_{1_x} & F_{1_y} \\ F_{2_x} & F_{2_y} \end{bmatrix} = \begin{bmatrix} -x & y/3 \\ 1/2 - 2x/3 & -2y/3 \end{bmatrix}$$

and

$$J^{-1}(\mathbf{F}(x,y)) = \frac{1}{-\frac{y}{6} + \frac{8xy}{9}} \begin{bmatrix} -\frac{2y}{3} & -\frac{y}{3} \\ -\frac{1}{2} + \frac{2x}{3} & -x \end{bmatrix}$$

Indicating $J^{-1}(\mathbf{F}(x,y))$ exists for all points except those on the x-axis and the ones on the line $x = \frac{3}{16}$. Thus, Newton–Raphson sequence for a system of equations takes the form

$$\mathbf{x}^{k+1} = \begin{bmatrix} x^{(k+1)} \\ y^{(k+1)} \end{bmatrix} = \mathbf{x}^k - J^{-1}(\mathbf{F}(\mathbf{x}^k))\mathbf{F}(\mathbf{x}^k)$$

$$= \begin{bmatrix} x^{(k)} \\ y^{(k)} \end{bmatrix} - \begin{bmatrix} -\frac{\frac{2y^{(k)}}{3}}{-\frac{y^{(k)}}{6} + \frac{8x^{(k)}y^{(k)}}{9}} & -\frac{\frac{y^{(k)}}{3}}{-\frac{y^{(k)}}{6} + \frac{8x^{(k)}y^{(k)}}{9}} \\ -\frac{-\frac{1}{2} + \frac{2x^{(k)}}{3}}{-\frac{y^{(k)}}{6} + \frac{8x^{(k)}y^{(k)}}{9}} & -\frac{x^{(k)}}{-\frac{y^{(k)}}{6} + \frac{8x^{(k)}y^{(k)}}{9}} \end{bmatrix}$$

$$\times \begin{bmatrix} \frac{-6[x^{(k)}]^2 + 2[y^{(k)}]^2 + 1}{12} \\ \frac{3x^{(k)} - 2[x^{(k)}]^2 - 2[y^{(k)}]^2 + 3}{6} \end{bmatrix}$$

Numerical results for this example are shown in Table 3.7.

The question that again arises is what if the Jacobian of $\mathbf{F}$ is not known, but only values of $\mathbf{F}(\mathbf{x})$ are known at discrete number of points? In that case, like the scalar case, $\frac{\partial f_i(\mathbf{x})}{\partial x_j}$ are replaced by their scant approximation of derivatives to find zeros of $\mathbf{F}(\mathbf{x})$.

Table 3.7. Comparing Newton–Raphson and Scant procedures.

#	Newton–Raphson method		Scant procedure	
n	$x(n)$	$y(n)$	$x(n)$	$y(n)$
0	3.0000	3.0000	3.0000	3.0000
1	1.6889	1.9833	2.9900	2.9900
2	1.1164	1.5604	1.6866	1.9812
3	0.9400	1.4395	1.2886	1.6844
4	0.9193	1.4263	1.0281	1.4987
5	0.9190	1.4262	0.9398	1.4394
6	0.9190	1.4262	0.9205	1.4270
7	0.9190	1.4262	0.9191	1.4262
8	0.9190	1.4262	0.9190	1.4262
9	0.9190	1.4262	0.9190	1.4262

3.6 Scant Method for a System of Equations

This method uses the following procedure to develop the sequence $\{\mathbf{x}^{(k)}\}$ that converges to the zeros of $\mathbf{F}(\mathbf{x})$ if starting point $\mathbf{x}^{(0)}$ is close enough, using following iteration:

$$\mathbf{x}^{(k+1)} = \mathbf{x}^{(k)} - \tilde{J}^{-1}(\mathbf{F}((\mathbf{k})))\mathbf{F}(\mathbf{x}^{(k)})$$

where elements of $\tilde{J}(\mathbf{F}(\mathbf{x}^{(k)}))$ are defined as

$$\tilde{J}_{ij}(\mathbf{F}(\mathbf{x}^{(k)})) = \frac{f_i(x_1^{(k)}, x_2^{(k)}, \ldots, x_j^{(k)}, \ldots, x_N^{(k)})}{x_j^{(k)} - x_j^{(k-1)}}$$

$$- \frac{f_i(x_1^{(k)}, x_2^{(k)}, \ldots, x_j^{(k-1)}, \ldots, x_N^{(k)})}{x_j^{(k)} - x_j^{(k-1)}}$$

Just like its scalar counterpart, for system of equations to start the Scant method, one needs to choose two starting points, $\mathbf{x}^{(0)}$ and $\mathbf{x}^{(1)}$. Numerical results for finding a zero of

$$\mathbf{F}(\mathbf{x}) = \begin{bmatrix} (-6x^2 + 2y^2 + 1)/12 \\ (3x - 2x^2 - 2y^2 + 3)/6 \end{bmatrix}$$

via secant method are also presented in Table 3.7. As expected, similar to the scalar case, Table 3.7 shows that Newton–Raphson method converges faster than Scant method.

MATLAB codes for Newton–Raphson and Scant procedures (Table 3.7)

```
% inputting initial information
xx(1) = 3; yy(1) = 3; xv(1) = xx(1); yv(1) = yy(1); xv(2) =
2.99; yv(2) = 2.99; Q = 9;
% finding Newton-Raphson & Scant sequences
for k = 1 : Q
ff = [(-6 * (xx(k))^2 + 2 * (yy(k))^2 + 1)/12(3 * xx(k) - 2 * (xx(k))^2 -
2 * (yy(k))^2 + 3)/6];   gj(1, 1) = -2 * yy(k)/3; gj(1, 2) = -yy(k)/3;
gj(2, 1) = -1/2 + 2 * xx(k)/3; gj(2, 2) = -xx(k);
fj = (1/(-yy(k)/6 + 8 * xx(k) * yy(k)/9)) * gj * ff;
xx(k + 1) = xx(k) - fj(1); yy(k + 1) = yy(k) - fj(2); end
for k = 2 : Q
fv = [(-6 * (xv(k))^2 + 2 * (yv(k))^2 + 1)/12(3 * xv(k) - 2 * (xv(k))^2 -
2 * (yv(k))^2 + 3)/6]; vj(1, 1) = ((-6 * (xv(k))^2 + 2 * (yv(k))^2 + 1)/12 -
(-6 * (xv(k-1))^2 + 2 * (yv(k))^2 + 1)/12)/(xv(k) - xv(k-1)); vj(1, 2) =
((-6 * (xv(k))^2 + 2 * (yv(k))^2 + 1)/12 - (-6 * (xv(k))^2 + 2 * (yv(k -
1))^2 + 1)/12)/(yv(k) - yv(k-1)); vj(2, 1) = ((3 * xv(k) - 2 * (xv(k))^2 -
2 * (yv(k))^2 + 3)/6 - (3 * xv(k-1) - 2 * (xv(k-1))^2 - 2 * (yv(k))^2 +
3)/6)/(xv(k) - xv(k-1)); vj(2, 2) = ((3 * xv(k) - 2 * (xv(k))^2 -
2 * (yv(k))^2 + 3)/6 - (3 * xv(k) - 2 * (xv(k))^2 - 2 * (yv(k-1))^2 +
3)/6)/(yv(k) - yv(k-1));
uj = inv(vj); pj = uj * fv;
xv(k + 1) = xv(k) - pj(1); yv(k + 1) = yv(k) - pj(2);
end [xx' yy' xv' yv']
```

Exercises

3.1. Use Theorem 3.1 to show that

$$g(x) = \frac{e^{-x^2}}{2}$$

defined over $x \in [-1, 1]$ has a unique fixed point. Apply fixed point iteration method to find it within 10^{-5} accuracy. How many iterations were needed to achieve the desired accuracy?

3.2. Write a MATLAB code that applies bisection method to find all roots of

$$f(x) = e^{-x^2} - 2x = 0, \quad \text{for } x \in [-1, 1]$$

within 10^{-5} accuracy. How many iterations were needed to achieve the desired accuracy?

3.3. Write a MATLAB code that applies the Newton–Raphson method to find all roots of

$$f(x) = e^{-x} - x^2 = 0, \quad \text{for } x \in [-1, 1]$$

within 10^{-5} accuracy. How many iterations were needed to achieve the desired accuracy?

3.4. If $f(x)$ has a root α with multiplicity m, prove that for any $m > 1$, the following modified Newton–Raphson iteration method

$$x_{n+1} = x_n - \frac{f'(x_n) f(x_n)}{f'(x_n)^2 - f(x_n) f''(x_n)}$$

converges quadratically provided starting value x_0 is close enough to α.

Hint: Prove that $g(x) = \frac{f(x)}{f'(x)}$ has a simple root at $x = \alpha$, by making use of the equation

$$f(x) = (x - \alpha)^m q(x), \quad \text{with } q(\alpha) \neq 0.$$

3.5. Prove the following iterative procedure convergences quadratically, when $f(x)$ has a root α with multiplicity $m > 1$.

$$x_{n+1} = x_n - \frac{m f(x_n)}{f'(x_n)}$$

Hint: Justify that $f(x)$ needs to have the following representation: $f(x) = (x - \alpha)^m q(x)$ with $q(\alpha) \neq 0$. Substitute this form of $f(x)$ in the above procedure to conclude the sequence

$$\left\{ \frac{|x_{n+1} - \alpha|}{|x_n - \alpha|^2} \right\}$$

converges to a finite positive constant.

3.6. (a) Use the modified Newton–Raphson iteration method given in Problem 3.4, to find the zero of

$$f(x) = (x^2 - 6x + 9)\ln(x)$$

which is not equal to one. Let $x_0 = 4$ be the starting point and stop the iteration when $|f(x_n)| \leq 10^{-6}$.

(b) Same as part (a) except use the standard Newton–Raphson iteration method to find the zero of $f(x)$.

(c) Compare part (a) and (b) results to see which procedure converges faster and discuss their advantages and the disadvantages.

3.7. To remedy the shortcoming of the procedure presented in Problem 3.5, prove that multiplicity of root α can be estimated by the equation

$$m \approx \frac{x_{n-1} - x_{n-2}}{2x_{n-1} - x_{n-2} - x_n}$$

where the sequence $\{x_n\}$ is generated via the standard Newton–Raphson method.

Hint: Show that

$$\frac{x_n - x_{n-1}}{(x_{n-1} - \alpha)} \approx -\frac{1}{m}, \text{ as } x_n \to \alpha.$$

3.8. (a) Same as Problem 3.6 except apply the modified Newton–Raphson procedure presented in Problem 3.5 to see how many iterations it takes to achieve the desired accuracy. To estimate m, use the equation given in Problem 3.7.

(b) Discuss the advantages and disadvantages of this modified Newton–Raphson method.

3.9. (a) Write a MATLAB code for the Secant method to find all roots of

$$f(x) = e^{-x} - x^2 = 0, \quad \text{for } x \in [-1, 1]$$

within 10^{-5} accuracy.

(b) How many iterations were needed to achieve the desired accuracy?

3.10. Write a MATLAB code that applies to the Jacobi's iterative method for finding fixed point of

$$\mathbf{F}(\mathbf{x}) = \frac{1}{9} \begin{bmatrix} x^2 + 2y + z + 1 \\ x + y^2 + 2z + 2 \\ 2x + y + z^2 + 1 \end{bmatrix}$$

where $\mathbf{x} = \begin{bmatrix} x \\ y \\ z \end{bmatrix}$. Let $\mathbf{x}^{(0)} = \mathbf{0}$. Terminate when $\|\mathbf{x}^{(k+1)} - \mathbf{x}^{(k)}\|_\infty \leq 10^{-5}$.

How many iterations were needed to achieve the desired accuracy?

3.11. Same as Problem 3.10, except apply Gauss–Seidel procedure to find the fixed point of $\mathbf{F}(\mathbf{x})$, and find number of iterations needed to achieve an accuracy of 10^{-5}.

3.12. (a) Let the vector valued function $\mathbf{F}(\mathbf{X})$ satisfy Lipschitz condition

$$\|\mathbf{F}(\mathbf{X}) - \mathbf{F}(\mathbf{Y})\|_\infty \leq L\|\mathbf{X} - \mathbf{Y}\|_\infty$$

Furthermore, assume the Lipschitz constant $L \in (0,1)$. Prove that

$$\|\vec{\alpha} - \mathbf{X}^{(k)}\|_\infty \leq \frac{L^k}{1-L}\|\mathbf{X}^{(0)} - \mathbf{X}^{(1)}\|_\infty$$

where $\mathbf{X}^{(k)} = \mathbf{F}(\mathbf{X}^{(k-1)})$, and $\vec{\alpha} = \mathbf{F}(\vec{\alpha})$.
(b) Prove that the defined sequence $\{\mathbf{X}^{(k)}\}$ is convergent.

3.13. Let

$$\mathbf{F}(\mathbf{X}) = \begin{bmatrix} -0.4x_1 + 0.2x_2 + 0.3x_3 + 1 \\ 0.1x_1 - 0.5x_2 + 0.3x_3 + 2 \\ 0.2x_1 + 0.1x_2 - 0.6x_3 + 3 \end{bmatrix}, \quad \text{where } \mathbf{X} = \begin{bmatrix} x_1 \\ x_2 \\ x_3 \end{bmatrix}$$

(a) Show that $\mathbf{F}(\mathbf{X})$ satisfies Lipschitz condition and find its Lipschitz constant L.
(b) Compute the sequence $\mathbf{X}^{(k+1)} = \mathbf{F}(\mathbf{X}^{(k)})$, starting with $\mathbf{X}^{(0)} = \mathbf{0}$.
(c) Will this sequence converge? If yes, terminate the iteration when error in estimating the fixed point $\vec{\alpha}$ becomes less than or equal to 10^{-5}.

3.14. Find a root of

$$\mathbf{F(x)} = \begin{bmatrix} x^2 - 8x + 2y + z + 2 \\ x + y^2 - 10y + 2z + 3 \\ 2x + y + z^2 - 12z + 4 \end{bmatrix} = \mathbf{0}$$

via Newton–Raphson method starting with $\mathbf{x}^{(0)} = \mathbf{0}$. Terminate your computer program when $\|\mathbf{x}^{(k+1)} - \mathbf{x}^{(k)}\|_\infty \leq 10^{-5}$. How many iterations were needed to achieve the desired accuracy?

3.15. Same as Problem 3.14, except apply Scant method to find the root of $\mathbf{F(x)}$. How many iterations were needed to achieve the desired accuracy, $\|\mathbf{x}^{(k+1)} - \mathbf{x}^{(k)}\|_\infty \leq 10^{-5}$?

Chapter 4

Polynomial Approximation and Interpolation

As we saw in the previous chapter for some studies, information about the rate of change of a function $f(x)$ is required, but its analytical form is not known and only numerical values of the function at some discrete points are available. Thus, many studies have been carried out to approximate/interpolate the function $f(x)$ from its values $y_i = f(x_i)$ available at $N + 1$ distinct discrete values of x_i, $i = 0, 1, 2\ldots, N$. One approach is the polynomial interpolation, where $N + 1$ data points $(x_i, y_i), i = 0, 1, 2, \ldots, N$ are given and we are to find a polynomial $P_N(x)$ that goes through all these $N + 1$ points. In other words, find the coefficients a_i such that

$$P_N(x_i) = a_N x_i^N + a_{N-1} x_i^{N-1} + \cdots + a_1 x_i + a_0 = y_i, \ n = 0, 1, 2, \ldots N$$

Such a requirement leads to the following system of linear equations with the coefficients $a_0, a_1, a_2, \ldots, a_N$ as the unknowns.

$$A \begin{pmatrix} a_0 \\ \vdots \\ a_N \end{pmatrix} = \begin{pmatrix} y_0 \\ \vdots \\ y_N \end{pmatrix}, \ A = \begin{pmatrix} 1 & x_0 & x_0^2 & \cdots & x_0^N \\ 1 & x_1 & x_1^2 & \cdots & x_1^N \\ \vdots & \vdots & \vdots & \vdots & \vdots \\ 1 & x_N & x_N^2 & . & x_N^N \end{pmatrix}$$

One could find the coefficients a_0, $a_1, \ldots, a_N$ by applying Gaussian elimination and back substitution procedure. But the solution to this problem is already known as the Lagrange interpolating polynomials.

4.1 Lagrange Interpolating Polynomials

The Lagrange polynomial is defined by the following equation:

$$P_N(x) = \sum_{k=0}^{N} y_k L_{N,k}(x), \text{ where } L_{N,k}(x) = \prod_{\substack{j=0 \\ j \neq k}}^{N} \frac{(x - x_j)}{(x_k - x_j)}$$

To show $P_N(x)$ is the desired solution, we note that $L_{N,k}(x_i) = \delta_{i,k}$. Thus, $P_N(x_i) = y_i$, for $i = 0, 1, 2, \ldots N$. In other words, $P_N(x)$ goes through all the nodes as required. Furthermore, this solution is unique, since if there exists another solution to the above system of equations, i.e. $Q_N(x)$ then

$$D_N(x) = P_N(x) - Q_N(x)$$

will be a polynomial of degree N with $D_N(x_i) = 0$, $i = 0, 1, 2 \ldots, N$, since by definition

$$D_N(x_i) = P_N(x_i) - Q_N(x_i) = y_i - y_i = 0, \quad \text{for } i = 0, 1, 2, \ldots N$$

In other words, $D_N(x)$ has $N + 1$ zeros, but since a nontrivial polynomial of degree N, cannot have more than N zeros, hence $D_N(x)$ must be identically zero for all x. Implying the found solution $P_N(x)$ is unique.

Example. Given two data point (x_0, y_0) & (x_1, y_1), associated with $N = 1$, find polynomial that goes through these two points. In other words

$$P_1(x) = a_1 x + a_0 \text{ find } a_0 \text{ & } a_1 \ni a_1 x_0 + a_0 = y_0 \text{ & } a_1 x_1 + a_0 = y_1$$

Using the Lagrange formula one finds

$$L_{1,0}(x) = \prod_{\substack{j=0,1 \\ j \neq 0}}^{N=1} \frac{(x - x_j)}{(x_0 - x_j)} = \frac{x - x_1}{x_0 - x_1},$$

$$L_{1,1}(x) = \prod_{\substack{j=0,1 \\ j \neq 1}}^{N=1} \frac{(x - x_j)}{(x_1 - x_j)} = \frac{x - x_0}{x_1 - x_0}$$

and substitute in $P_N(x) = \sum_{k=0}^{N} y_k L_{N,k}(x)$, to find

$$P_1(x) = +y_0 L_{10} + y_1 L_{11} = y_0 \frac{x - x_1}{x_0 - x_1} + y_1 \frac{x - x_0}{x_1 - x_0}$$

As it can be seen from the above $P_1(x_0) = y_0$, & $P_1(x_1) = y_1$. In other words, defined $P_1(x)$ does go through the desired points, and it is well known that there is only one unique straight line that goes through any two points.

Theorem 4.1. *If* $f \in C^{N+1}[a,b]$ *&* $x_0, \ldots, x_N \in [a,b]$ *are* $N+1$ *distinct nodes of* $f(x)$*, and* $P_N(x)$ *is the Lagrange interpolating polynomial satisfying* $P_N(x_i) = f(x_i) \; \forall \; x_i$ *with* $n = 0, 1, 2, \ldots N$*, then exists a* $\xi \in (a,b)$*, such that the difference between* $P_N(x)$ *&* $f(x)$ *is*

$$R_N(x) = f(x) - P_N(x) = \frac{(x - x_0)(x - x_1)\ldots(x - x_N)}{(N+1)!} f^{(N+1)}(\xi)$$

Exercise 4.2 provides the details on how to prove Theorem 4.1.

It may appear that by making N larger, the error/remainder $R_N(x)$ will become smaller, however in general this is not the case. It turns out that for larger N the interpolating polynomial $P_N(x)$ accurately approximates $f(x)$ for values of x near the midpoint $(a+b)/2$, but when x is near the end points a or b the error $R_N(x)$ may exhibit large oscillations. See Fig. 4.1.

It has also been noticed that values of $R_N(x)$ depends on the values of nodal points used to sample $f(x)$. The popular sampling strategy of equally spaced nodes turns out not to minimize variation of $R_N(x)$.

4.2 Chebyshev Polynomials

Since the error as given by Theorem 4.1 is

$$R_N(x) = \frac{(x - x_0)(x - x_1) \cdots (x - x_N)}{(N+1)!} f^{(N+1)}(\xi)$$

Chebyshev studied how to choose $x_0, x_1, \ldots, x_N$ in order to minimize the variations in $R_N(x)$ from its ideal value of zero. In order for P_N to be the "best" approximation of $f(x)$ for $x \in [-1,1]$, and since

$f^{(N+1)}(\xi)$ is not directly influenced by the sampling points x_k, he concentrated on the $N+1$ degree polynomial

$$\phi(x) = (x - x_0)(x - x_1) \cdots (x - x_N)$$

that can minimize $R_N(x)$ over $[-1, 1]$, by an appropriate selection of the interpolating points x_k for $k = 0, 1, \ldots N$. To achieve this, Chebyshev changed the variable x to $\theta \ni x = \cos \theta$, and defined

$$T_n(x) = \cos(n\theta) = \cos(n \cos^{-1} x)$$

Although $T_n(x)$ does not look like a polynomial, but as we shall see it will become intimately related to $\phi(x)$. To see this, we apply the definition of Chebyshev polynomials $T_n(x)$ to find

$$T_0(x) = \cos(0) = 1, T_1(x) = \cos \theta = \cos(\cos^{-1} x) = x$$
$$T_2 = \cos 2\theta = 2(\cos \theta)(\cos \theta) - 1 = 2\cos(\cos^{-1} x)T_1 - T_0$$
$$= 2xT_1 - T_0$$
$$T_{N+1}(x) = \cos(N\theta + \theta) = \cos(N\theta)\cos(\theta) - \sin(N\theta)\sin(\theta)$$
$$T_{N-1}(x) = \cos(N\theta - \theta) = \cos(N\theta)\cos(\theta) + \sin(N\theta)\sin(\theta)$$
$$\text{Thus, } T_{N+1}(x) + T_{N-1}(x) = 2xT_N(x)$$

From the above equations, the following recursion is found:

$$T_{N+1}(x) = 2xT_N(x) - T_{N-1}(x), \quad \text{for } -1 \le x \le 1$$

Above examples and recursion relation indicate $T_{N+1}(x)$ is a polynomial of degree $N+1$ with leading coefficient 2^N. This finding motives the definition of Monic Chebyshev polynomials.

$$\tilde{T}_{N+1} = \frac{T_{N+1}}{2^N}$$

Monic polynomials are defined to be polynomials with leading coefficient equal to one.

     Comparing $\tilde{T}_n$ to all monic polynomials of degree n, which we denote as a family of monic polynomials $\tilde{P}_n$, Chebyshev arrived at the following result.

Theorem 4.2. *Monic Chebyshev polynomials $\tilde{T}_n$ of degree $n \geq 1$ satisfy*

$$\frac{1}{2^{n-1}} = \max_{x \in [-1,1]} |\tilde{T}_n(x)| \leq |\tilde{P}_n(x)| \quad \forall \; \tilde{P}_n \in \tilde{\mathcal{P}}_n$$

with equality being the case when zeros of $\tilde{P}_n(x)$ is the same as zeros of $\tilde{T}_n(x)$. For example for $n = N + 1$, $T_{N+1}(x)$ has simple zeros at

$$\theta_k = \frac{(2k+1)\pi}{2N+2} \quad for \; k = 0, \, 1, \ldots N$$

since $T_{N+1}(x_k) = \cos((N+1)\theta_k) = \cos((2k+1)\frac{\pi}{2}) = 0$. Thus, if we take interpolation points x_k to be at the zeros of Chebyshev polynomials

$$x_k = \cos\left[\frac{(2k+1)\pi}{2N+2}\right], \quad k = 0, 1, 2, \ldots N$$

then

$$\phi(x) = (x - x_0)(x - x_1)\cdots(x - x_N) = \tilde{T}_{N+1}(x) = 2^{-N}T_{N+1}(x)$$

will minimize $|R_N(x)| = |\phi(x)|\frac{|f^{(N+1)}(\xi)|}{(N+1)}$ for $x \in [-1, 1]$. Furthermore, since the extreme values of T_{N+1} occur at

$$\theta_j = \frac{j\pi}{N+1}, \quad j = 0, 1, 2, \ldots, N+1$$

with $\max |T_{N+1}| = 1 \; \forall \; x \in [-1, 1]$, one finds

$$|R_N(x)| \leq \frac{1}{2^N(N+1)!} \max_{x \in [-1,1]} |f^{(N+1)}(x)|$$

In other words, using zeros of Chebyshev polynomials as nodes for interpolation of a function $f(x)$ will provide us with an interpolating polynomial that has the smallest deviation from $f(x)$.

Figure 4.1 shows results of interpolating the function $\exp(-12x^2)$ by equally spaced nodes and by Chebyshev nodes for $N = 10$. It shows for points near the mid-point, the interpolations are satisfactory for both types of nodes, but for values of x near the endpoints considerable deviation from $f(x)$ can set in, especially for equally

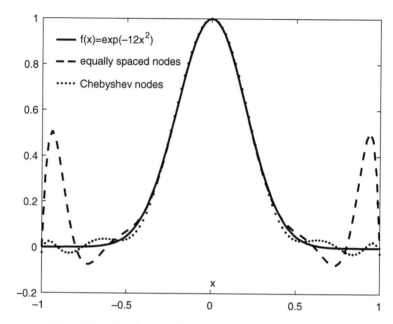

Fig. 4.1. P_{10} for equally spaced and Chebyshev nodes.

spaced nodes. This oscillation is called Runge phenomenon which becomes even more pronounced for larger N.

MATLAB code for generating Fig. 4.1.

```
% inputting initial data
M=100; b=1; a=-b; dx = (b-a)/M; M1=M+1; N=10; N1=N+1;
dx1 = (b-a)/N;
% finding (v, u) equally spaced nodes &  (w, z) Chebyshev nodes
for m=1:N1
v(m)=a+(m-1)*dx1; u(m)=exp(-12*(v(m))^2);
arg=((2*m-1)*π)/(2*N1); w(m)=cos(arg);
z(m)=exp(-12*(w(m))^2); end;
for n=1:M1
x(n)=a+(n-1)*dx; y(n)=exp(-12*(x(n))^2); end;
% calculating Lagrange polynomial for selected nodes
for n=1:M1
yy=0; yy1=0;
for k=1:N1
```

```
LD(k)=1; LN(k)=1; CD(k)=1; CN(k)=1;
for j=1:N1
if ( j==k ) LMD=1; LMN=1; CMD=1; CMN=1; else
LMD=v(k)-v(j);  LMN=x(n)-v(j);  CMD=w(k)-w(j);  CMN=x(n)-
w(j); end;
LD(k)=LD(k)*LMD; LN(k)=LN(k)*LMN; CD(k)=CD(k)*CMD;
CN(k)=CN(k)*CMN; end;
yy=yy+u(k)*LN(k)/LD(k); yy1=yy1+z(k)*CN(k)/CD(k); end;
y1(n)=yy; y2(n)=yy1; end;
plot(x,y,'k',x,y2,'k.', x,y1,'k-','linewidth',2.0)
```

For simplicity, Chebyshev theory was presented for $x \in [-1, 1]$. To apply it to an arbitrary interval $[a, b]$, when $f(x)$ is defined over $a \leq x \leq b$, one needs to transform the variable x to a new variable $y \in [-1, 1]$, by defining:

$$x = \frac{a + b}{2} + \frac{(b - a)}{2}y, \quad \text{for } -1 \leq y \leq 1$$

This transformation maps $[-1, 1]$ onto $[a, b]$. Thus, Chebyshev nodes for $x \in [a, b]$ are found via the following equations:

$$x_k = \frac{a + b}{2} + \frac{(b - a)}{2}y_k, \quad \text{where } y_k = \cos\left(\frac{(2k + 1)\pi}{2N + 2}\right),$$

$$k = 0, 1, 2, \ldots, N$$

Above nodal points gives best locations of x_k for sampling a function defined over the interval $[a, b]$. But collecting data at the Chebyshev nodes x_k, $k = 0, 1, 2, \ldots N$, may not always be feasible. Thus, based on the above findings, and to avoid onset of instability, one needs to make degree of the interpolating polynomials small. However, if N is too small then Lagrange interpolating function $P_N(x)$ may not be able to approximate $f(x)$ in a satisfactory manner. Also, as already mentioned, not always data associated with Chebyshev nodes are available. These considerations motivate the need to develop other interpolating procedures. We will discuss two methods that enable one to approximate a function using polynomials with lower degrees.

4.3   Spline Interpolation

The desire to keep interpolating polynomial to a low degree, and pass through all the nodal points, leads to defining a continuous function $S(x)$ which is a low degree polynomial over subintervals $[x_k, x_{k+1}], k = 0, 1, 2, \ldots, (N-1)$, where data is given as $f(x_k) = y_k$ and $S(x)$ needs to pass through all the points (x_k, y_k), $k = 0, 1, \ldots, N$. A commonly used spline function $S(x)$ is a piecewise cubic polynomial defined over subintervals $[x_k, x_{k+1}]$, $x_k < x_{k+1}$, when $k = 0, 1, \ldots, (N-1)$. Furthermore, it is required that first and second derivatives of $S(x)$ to be continuous over $[x_0, x_N]$. Such a spline function is called Cubic Splines. The spline function $S(x)$ is usually represented as:

$$S(x) \equiv s_k(x) \text{ for } x_k \le x \le x_{k+1}, \quad k = 0, 1, 2, \ldots, (N-1), \text{ where}$$

$$s_k(x) = y_k + \alpha_k(x - x_k) + \beta_k(x - x_k)^2 + \gamma(x - x_k)^3,$$

$$k = 0, 1, 2, \ldots (N-1)$$

By its definition and continuity requirement, it follows:

$$s_j(x_j) = y_j, \ j = 0, 1, 2, \ldots, N - 1 \quad \text{and}$$

$$s_{k-1}(x_k) = y_k, \ k = 1, 2, \ldots, N$$

Also the definition of $s_k(x)$ leads to $s_k''(x)$ being a linear function of x. Thus, it takes the form:

$$s_k''(x) = \frac{(x - x_k)}{x_{k+1} - x_k} s_k''(x_{k+1}) + \frac{(x - x_{k+1})}{x_k - x_{k+1}} s_k''(x_k),$$

$$k = 0, 1, 2, \ldots, (N-1)$$

Integrating the above equation twice, one finds:

$$s_k(x) = \frac{(x - x_k)^3}{6c_k} r_{k+1} - \frac{(x - x_{k+1})^3}{6c_k} r_k$$

$$+ \eta_k(x_{k+1} - x) + \xi_k(x - x_k)$$

where

$$r_k = s_k''(x_k), \quad c_k = x_{k+1} - x_k$$

and η_k & ξ_k are associated with the constants of integration over subinterval $[x_k, x_{k+1}]$. To find η_k & ξ_k we use following continuity relations satisfied by $S(x)$ and its two derivatives.

$$s_k''(x_{k+1}) = s_{k+1}''(x_{k+1}) = r_{k+1}$$

$$s_k(x_k) = y_k = \frac{c_k^2}{6}r_k + \eta_k c_k,$$

$$s_k(x_{k+1}) = s_{k+1}(x_{k+1}) = y_{k+1} = \frac{c_k^2}{6}r_{k+1} + \xi_k c_k$$

Hence,

$$s_k(x) = \frac{r_k}{6c_k}(x_{k+1} - x)^3 + \frac{r_{k+1}}{6c_k}(x - x_k)^3 + \left(\frac{y_k}{c_k} - \frac{r_k c_k}{6}\right)(x_{k+1} - x)$$

$$+ \left(\frac{y_{k+1}}{c_k} - \frac{r_{k+1}c_k}{6}\right)(x - x_k),$$

$$s_k'(x) = -\frac{r_k}{2c_k}(x_{k+1} - x)^2 + \frac{r_{k+1}}{2c_k}(x - x_k)^2 + d_k - \frac{(r_{k+1} - r_k)c_k}{6},$$

where $d_k = \frac{y_{k+1}-y_k}{c_k}$ and $k = 0, 1, 2, \ldots, (N-1)$.

Applying continuity of $S'(x)$ at nodal points, that is $s_{k-1}'(x_k) = s_k'(x_k)$ one finds

$$c_{k-1}r_{k-1} + 2(c_{k-1} + c_k)r_k + c_k r_{k+1} = g_k, \quad k = 1, 2, \ldots, (N-1)$$

where $g_k = 6(d_k - d_{k-1})$.

The above system of equations results in $N - 1$ equations with $N + 1$ unknowns: $r_0, \ldots, r_k, \ldots, r_N$. To remedy this lack of information and uniqueness, two more conditions are needed to be introduced. This requirement results in different choices for finding the cubic spline coefficients. We present only two commonly used cubic splines. If one is to choose $r_0 = r_N = 0$, the resulting solution is called **Natural spline**. On the other hand if values of $S'(x_0)$ and $S'(x_N)$ are assumed to be known, the resulting solution is called **Clamped spline**.

For natural spline where $r_0 = r_N = 0$, the above set of equations for spline function takes the form

$$a_1 r_1 + c_1 r_2 = g_1, \qquad \text{for } k = 1$$

$$c_{k-1}r_{k-1} + a_k r_k + c_k r_{k+1} = g_k, \qquad \text{for } k = 2, \ldots, N-2$$

$$c_{N-2}r_{N-2} + a_{N-1}r_{N-1} = g_{N-1} \qquad \text{for } k = N-1$$

where $a_k = 2(c_k + c_{k-1})$ with the unknowns being r_k for $k = 1, 2, 3, \ldots, (N-1)$. Writing this system of equations in a vector form

$$A\mathbf{r} = \mathbf{g}$$

with

$$
A = \begin{pmatrix}
a_1 & c_1 & . & & . & \\
b_2 & a_2 & c_2 & & . & \\
. & . & . & . & & \\
. & . & . & . & & \\
. & . & . & b_{N-1} & a_{N-1}
\end{pmatrix}, \quad
\mathbf{r} = \begin{pmatrix}
r_1 \\
r_2 \\
. \\
r_{N-1}
\end{pmatrix}, \quad
\mathbf{g} = \begin{pmatrix}
g_1 \\
g_2 \\
. \\
g_{N-1}
\end{pmatrix}
$$

and $b_n = c_{n-1}$. The matrix A is a tridiagonal matrix and its inverse exists since it is a diagonally dominated matrix

$$|a_1| > |c_1| > 0$$
$$|a_k| \geq |b_k| + |c_k|$$
$$|a_{N-1}| > |b_{N-1}| > 0$$

Thus,

$$\mathbf{r} = A^{-1}\mathbf{g}$$

Having calculated r_k for $k = 1, 2, \ldots, (N-1)$ and assuming $r_0 = r_N = 0$, the natural spline $s_k(x)$ is found for a given data set.

For Clamped cubic spline, one needs to find $r_k, k = 0, 1, 2, \ldots, N$, since r_0 and r_N are not given, one makes use of specified values $S'(x_0)$ and $S'(x_N)$ to find r_k. That is applying the relation

$$s_k'(x) = -\frac{r_k}{2c_k}(x_{k+1} - x)^2 + \frac{r_{k+1}}{2c_k}(x - x_k)^2 + d_k - \frac{(r_{k+1} - r_k)c_k}{6}$$

From the above for $k = 0$, one finds

$$S'(x_0) = -\frac{r_0}{2}c_0 + d_0 - \frac{(r_1 - r_0)c_0}{6}$$

or

$$r_0 + \frac{r_1}{2} = \frac{3}{c_0}(d_0 - S'(x_0))$$

Likewise for $k = N - 1$, one finds

$$r_N + \frac{r_{N-1}}{2} = \frac{3}{c_{N-1}}(S'(x_N) - d_{N-1})$$

Including these two equations along with the other $N - 1$ equations found for cubic spline interpolation of the data, provide us with $N+1$ equations and $N + 1$ unknowns. Writing this set of equations in a vector form

$$\tilde{A}\tilde{\mathbf{r}} = \mathbf{w}$$

$$\tilde{A} = \begin{pmatrix} 1 & 1/2 & . & . & . & & . & & . \\ b_1 & a_1 & c_1 & . & . & & . & & . \\ . & . & . & . & . & & . & & . \\ . & . & . & b_k & a_k & & c_k & & . \\ . & . & . & . & b_{N-1} & a_{N-1} & c_{N-1} \\ . & . & . & . & . & 1/2 & 1 \end{pmatrix}, \quad \tilde{\mathbf{r}} = \begin{pmatrix} r_0 \\ r_1 \\ . \\ r_k \\ . \\ r_N \end{pmatrix},$$

$$\mathbf{w} = \begin{pmatrix} w_0 \\ w_1 \\ . \\ w_k \\ . \\ w_N \end{pmatrix}$$

$w_k = g_k$, for $k = 1, 2, \ldots, (N - 1)$, $w_0 = \frac{3}{c_0}(d_0 - S'(x_0))$ and $w_N = \frac{3}{c_{N-1}}(S'(x_N) - d_{N-1})$.

Like the case of natural spline, $\tilde{A}$ is also diagonally dominated and its inverse exist. Thus,

$$\tilde{\mathbf{r}} = \tilde{A}^{-1}\mathbf{w}$$

Having obtained r_k for $k = 0, 1, 2, \ldots, N$. the clamped cubic spline formula for interpolation of a given data set is found.

Figure 4.2 shows the performance of the spline interpolation of $f(x) = \frac{1}{1+10x^2}$ for nodal points $x_k = -1 + \frac{2k}{N}$, $k = 0, 1, 2, \ldots, N$, using natural spline.

MATLAB code for generating Fig. 4.2.

```
% defining (sx, sy) nodes, & the interpolating data points
for q=1:11;
sx(q)=-1+(q-1)*0.2; sy(q)= 1/(1 + 10 * (sx(q))^2);
```

end;
% using MATLAB code to find spline function
sxx=-1:0.02:1; syy=spline(sx,sy,sxx);
plot(x,y,'k–',sxx,syy,'k.', sx,sy,'kO','linewidth',1.5)

Comparing Figs. 4.1 and 4.2 we notice that as expected the oscillation near the end of intervals called Runge phenomenon disappears in Fig. 4.2, since polynomials used for subinterval segments are only of order 3. For the example given in Fig. 4.2, agreement with its natural cubic spline interpolation result looks good, as compared to the result shown in Fig. 4.1, but one had to find the interpolating function over different segments of the interval.

4.4 Least Square Fitting of Data

There are situations where the general form of a function is known except for its dependence on some parameters which are to be found from a given set of data. For example let $f(x) = a_0 + a_1 x$ and we are to find coefficients a_0 & a_1 from data points (x_k, y_k) for $k = 1, 2, \ldots N$,

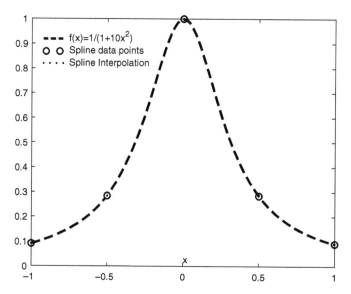

Fig. 4.2. Spline interpolation for equally spaced nodes.

where $N > 2$. Since a straight line is represented by a polynomial of degree one, interpolation approach presented so far will not be very helpful, since a straight line in general cannot go through all the given points. This problem necessitates the need to define what we mean by the "best" straight line that represents the given data points. The standard approach to this problem is to find the error or the difference between the selected straight line and the data points at every individual nodes

$$e_k = f(x_k) - y_k, \quad k = 1, 2, \ldots, N$$

Next select a desired norm to calculate totality of these errors and make them to be as small as possible. It turns out due to practical considerations, the norm most often used in this context is the l_2 norm, which is called the Least Square fit to the data.

$$E(f) = \sum_{k=1}^{N} e_k^2 = \sum_{k=1}^{N} (f(x_k) - y_k)^2$$

For our example with $f(x) = a_0 + a_1 x$, total error would be

$$E(a_0, a_1) = \sum_{k=1}^{N} (a_0 + a_1 x_k - y_k)^2$$

and the best least square straight line to the data would be the one that results in $E(a_0, a_1)$ to be as small as possible. In other words

$$\frac{\partial E(a_0, a_1)}{\partial a_0} = 0 = \sum_{k=1}^{N} 2(a_0 + a_1 x_k - y_k)$$

$$\frac{\partial E(a_0, a_1)}{\partial a_1} = 0 = \sum_{k=1}^{N} 2(a_0 + a_1 x_k - y_k) x_k$$

Simplify above two equations to find

$$a_0 N + a_1 \sum_{k=1}^{N} x_k = \sum_{k=1}^{N} y_k$$

$$a_0 \sum_{k=1}^{N} x_k + a_1 \sum_{k=1}^{N} x_k^2 = \sum_{k=1}^{N} y_k x_k$$

Equivalently,

$$Q\mathbf{a} = \mathbf{b}, \quad \text{where}$$

$$\mathbf{a} = \begin{pmatrix} a_0 \\ a_1 \end{pmatrix}, \quad \mathbf{b} = \begin{pmatrix} \sum_{k=1}^{N} y_k \\ \sum_{k=1}^{N} y_k x_k \end{pmatrix}, \quad Q = \begin{pmatrix} N & \sum_{k=1}^{N} x_k \\ \sum_{k=1}^{N} x_k & \sum_{k=1}^{N} x_k^2 \end{pmatrix}$$

From which a_0 & a_1 are found. The procedure can also be extended to a linear combination of several functions. That is

$$f(x) = \sum_{m=1}^{M} b_m f_m(x)$$

with $f(x)$ to be the best fit for N data points (x_k, y_k). The functions $f_m(x)$, $m = 1, 2, \ldots, M$ are given for $M \leq N$. The coefficients b_m are to be found using least square procedure. A common example of such a function $f(x)$ is a polynomial of degree $M - 1$, i.e.

$$f(x) = \sum_{m=1}^{M} b_m f_m(x), \quad \text{where } f_m(x) = x^{m-1}$$

Again the total error is defined as

$$E(b_1, \ldots, b_m, \ldots, b_M) = \sum_{k=1}^{N} (f(x_k) - y_k)^2$$

$$\frac{\partial E}{\partial b_m} = 2 \sum_{k=1}^{N} (f(x_k) - y_k) \frac{\partial f(x_k)}{\partial b_m}$$

$$= 2 \sum_{k=1}^{N} (f(x_k) - y_k) \frac{\partial}{\partial b_m} \sum_{j=1}^{M} b_j f_j(x_k)$$

$$= 2 \sum_{k=1}^{N} (f(x_k) - y_k) f_m(x_k) = 0, \quad m = 1, 2, \ldots, M$$

or

$$\sum_{k=1}^{N} f(x_k) f_m(x_k) = \sum_{k=1}^{N} y_k f_m(x_k)$$

$$\sum_{k=1}^{N} \left(\sum_{j=1}^{M} b_j f_j(x_k) \right) f_m(x_k) = \sum_{k=1}^{N} y_k f_m(x_k)$$

$$= \sum_{j=1}^{M} b_j \sum_{k=1}^{N} f_j(x_k) f_m(x_k)$$

Writing the above in a matrix form:

$$D\mathbf{b} = \mathbf{c}, \text{ where } \mathbf{b} = \begin{pmatrix} b_1 \\ \vdots \\ b_M \end{pmatrix}, \quad \mathbf{c} = \begin{pmatrix} \sum_{k=1}^{N} y_k f_1(x_k) \\ \vdots \\ \sum_{k=1}^{N} y_k f_M(x_k) \end{pmatrix}, \quad \text{and}$$

$$D = \begin{pmatrix} \sum_{k=1}^{N} f_1(x_k) f_1(x_k) & \sum_{k=1}^{N} f_1(x_k) f_2(x_k) & \cdots & \sum_{k=1}^{N} f_1(x_k) f_M(x_k) \\ \vdots & \vdots & \vdots & \vdots \\ \sum_{k=1}^{N} f_M(x_k) f_1(x_k) & \sum_{k=1}^{N} f_M(x_k) f_2(x_k) & \cdots & \sum_{k=1}^{N} f_M(x_k) f_M(x_k) \end{pmatrix}$$

If $\det D \neq 0$, the above equation will provide us with the least square fit to the data:

$$\mathbf{b} = D^{-1}\mathbf{c}$$

A natural extension of the above procedure is to see how to apply the method if the unknown coefficients do not appear in a linear manner. For example requiring $f(x) = ae^{bx}$ to best fit a given data set (x_k, y_k), $k = 1, 2, \ldots, N$. Following same procedure

$$E(a, b) = \sum_{k=1}^{N} (y_k - ae^{bx_k})^2$$

$$\frac{\partial E}{\partial a} = -2 \sum_{k=1}^{N} (y_k - ae^{bx_k}) e^{bx_k}$$

$$\frac{\partial E}{\partial b} = -2 \sum_{k=1}^{N} (y_k - ae^{bx_k})(ax_k) e^{bx_k}$$

The problem is now finding a & b from following set of nonlinear equations:

$$f_1(a, b) = a \sum_{k=1}^{N} e^{2bx_k} - \sum_{k=1}^{N} y_k e^{bx_k} = 0$$

$$f_2(a,b) = a \sum_{k=1}^{N} x_k e^{2bx_k} - \sum_{k=1}^{N} x_k y_k e^{bx_k} = 0$$

Thus, the best fit problem is reduced to finding zeros of

$$\mathbf{F}(a,b) = \begin{pmatrix} f_1(a,b) \\ f_2(a,b) \end{pmatrix}$$

The Newton–Raphson method could be used to find a & b. But this example shows the best fit procedure for nonlinear dependence of f on the parameters may not be as simple as for the case when $f(x)$ depends linearly on the unknown parameters. However, it turns out sometimes it is possible to transform a given function, so that the unknown coefficients appear linearly in the transformed function. For example, consider the function just studied: $y = ae^{bx}$. If this function is transformed using the log transform

$$\ln y = \ln a + bx = \tilde{y} = \tilde{a} + bx$$

where $\tilde{y} = \ln y$ and $\tilde{a} = \ln a$, then the problem is reduced to finding the best fit to the modified data by a straight line. Numerical results for such a fit is shown in Table 4.1.

Table 4.1 shows a set of data points (x_k, y_k) and quantities needed to be computed for finding best fit by $f(x) = ae^{bx}$. As mentioned, data is transformed by the log function and $\sum x_k$, $\sum \tilde{y}_k$, $\sum x_k^2$, & $\sum x_k \tilde{y}_k$ are computed. To find $\tilde{a}$ & b, quantities defined in equation $Q\mathbf{a} = \mathbf{b}$ for the given data need to be found.

Table 4.1. Needed quantities for finding least square fit to above data points (x_k, y_k).

n	x_n	y_n	$\tilde{y}_n$	x_n^2	$x_n \tilde{y}_n$
1	0	1.1000	0.0953	0	0
2	0.2000	1.5767	0.4553	0.0400	0.0911
3	0.4000	2.2599	0.8153	0.1600	0.3261
4	0.6000	3.2391	1.1753	0.3600	1.2282
6	1.0000	6.6546	1.8953	1.0000	1.8953
Sum	3	19.4731	5.9719	2.2	4.2459

That is

$$\mathbf{a} = \begin{pmatrix} \tilde{a} \\ b \end{pmatrix}, \quad \mathbf{b} = \begin{pmatrix} \sum_{k=1}^{N} \tilde{y}_k \\ \sum_{k=1}^{N} \tilde{y}_k x_k \end{pmatrix} = \begin{pmatrix} 5.9719 \\ 4.2459 \end{pmatrix} \text{ and }$$

$$Q = \begin{pmatrix} N & \sum_{k=1}^{N} x_k \\ \sum_{k=1}^{N} x_k & \sum_{k=1}^{N} x_k^2 \end{pmatrix} = \begin{pmatrix} 6 & 3 \\ 3 & 2.2 \end{pmatrix},$$

$$Q^{-1} = \begin{pmatrix} 0.5238 & -0.7143 \\ -0.7143 & 1.4286 \end{pmatrix}$$

$$\begin{pmatrix} \tilde{a} \\ b \end{pmatrix} = Q^{-1}\mathbf{b} = \begin{pmatrix} 0.0953 \\ 1.8000 \end{pmatrix}$$

Thus, $a = e^{\tilde{a}} = e^{0.0953} = 1.1$ & $b = 1.800$, resulting the least square fit to have the following form:

$$f(x) = 1.1e^{1.8x}$$

The least square fit to the data is shown in Fig. 4.3. It shows the function that fits the selected data points is exactly the same function that generated the data. However, if there were some data errors, the method would still have found a least square exponential fit to the data but may not have gone through all the actual data points without the measurement error.

MATLAB code for generating Fig. 4.3

```
% inputting initial information
M=50; b=1; a=0; dx=(b-a)/M; M1=M+1; N=6; sx=0; sy=0;
sz=0;sx2=0;sxz=0;
% computing nodal points
for k=1:N
x(k)=(k-1)*0.2; y(k)=1.1*exp(1.8*x(k)); z(k)=log(y(k));
x2(k)=x(k)*x(k);
xz(k)=x(k)*z(k); sx=sx+x(k); sx2=sx2+x2(k);
sz=sz+z(k); xz=sxz+x(k)*z(k);sy=sy+y(k); end
B=[sz;sxz]; Q=[N sx; sx sx2]; QI=inv(Q); A=QI*B; at=exp(A(1));
% computing the least square fit
```

for m=1:M1 xx(m)=(m-1)*dx; yy(m)=at*exp(xx(m)*A(2)); end
plot(xx,yy,'k–', x,y,'ko','LineWidth',2.0)

Other functions also could be transformed to become linearly dependent on modified version of the unknown parameters. For example,

$$y = f(x) = \frac{1}{b_1 + b_2 x^2}$$

which is not linearly dependent on the parameters b_1 & b_2, but equivalent equation

$$\tilde{y} = b_1 + b_2 x^2, \quad \text{where } \tilde{y} = \frac{1}{y}$$

is linearly dependent on the parameters. Thus, if one can transform data points (x_k, y_k) to $(x_k, \tilde{y}_k)$, $k = 1, 2, \ldots, N$, the problem becomes linear in the unknown coefficients b_1 & b_2 and the least square procedure can be applied without any major modification. As an

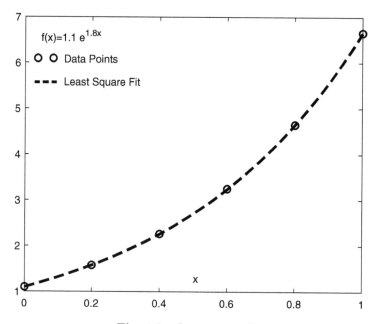

Fig. 4.3. Least square fit.

Table 4.2. Needed quantities for finding least square fit to
above data points (x_k, y_k).

n	x_n	y_n	$\tilde{y}_k$	x_n^2	x_n^4	$x_n^2\tilde{y}_n$
1	−1.0000	0.0909	11.0000	1.0000	1.0000	11.0000
2	−0.8000	0.1351	7.4000	0.6400	0.4096	4.7360
3	−0.6000	0.2174	4.6000	0.3600	0.1296	1.6560
4	−0.4000	0.3846	2.6000	0.1600	0.0256	0.4160
5	−0.2000	0.7143	1.4000	0.0400	0.0016	0.0560
6	0	1.0000	1.0000	0	0	0
7	0.2000	0.7143	1.4000	0.0400	0.0016	0.0560
8	0.4000	0.3846	2.6000	0.1600	0.0256	0.4160
9	0.6000	0.2174	4.6000	0.3600	0.1296	1.6560
10	0.8000	0.1351	7.4000	0.6400	0.4096	4.7360
11	1.0000	0.0909	11.0000	1.0000	1.0000	11.0000
Sum	0.0000	4.0847	55.0000	4.4000	3.1328	35.7280

example let us try fitting data used in generating for cubic spline
function as shown in Fig. 4.2, that is $f(x) = \frac{1}{1+10x^2}$. That is let
the generated data be fitted by $f(x) = \frac{1}{b_1+b_2x^2}$. Similar to other
examples, we need to compute quantities for the least square fit to
$\tilde{f}(x) = b_1 + b_2x^2 = \frac{1}{f(x)}$. In other words: $\tilde{f}(x) = \sum_{m=1}^{M} b_m f_m(x)$,
$f_1(x) = 1$, $f_2(x) = x^2$, with

$$\mathbf{c} = \begin{pmatrix} c_1 \\ c_2 \end{pmatrix} = \begin{pmatrix} \sum_{k=1}^{N} \tilde{y}_k \\ \sum_{k=1}^{N} \tilde{y}_k x_k^2 \end{pmatrix}, \quad D = \begin{pmatrix} \sum_{k=1}^{N} 1 & \sum_{k=1}^{N} x_k^2 \\ \sum_{k=1}^{N} x_k^2 & \sum_{k=1}^{N} x_k^4 \end{pmatrix}, \quad \text{and}$$

$$\mathbf{b} = \begin{pmatrix} b_1 \\ b_2 \end{pmatrix} = D^{-1}\mathbf{c}$$

Numerical results are shown in Table 4.2, and graph of its least square
approximation of the data given by $\frac{1}{1+10x^2}$ is shown in Fig. 4.4.

MATLAB code for generating Fig. 4.4:

```
% inputting initial information
M=50; b=1; a=-1; dx=(b-a)/M; M1=M+1; N=11;
sx=0; sy=0;sz=0;sq=0;sqz=0;sq2=0;
% finding nodes and needed coefficients
for k=1:N
```

x(k)=a+(k-1)*0.2; y(k)=1/(1+10*x(k)^2); z(k)=1/y(k); q(k)=x(k)*x(k);
q2(k)=q(k)*q(k); qz(k)=q(k)*z(k); sx=sx+x(k); sq=sq+q(k);
sq2=sq2+q2(k); sz=sz+z(k); sqz=sqz+q(k)*z(k);sy=sy+y(k); end
B=[sz;sqz]; Q=[N sq; sq sq2]; QI=inv(Q); D=QI*B;
d1= D(1); d2= D(2);
% finding the least square fit
for m=1:M1 xx(m)=a+(m-1)*dx; yy(m)=1/(d1+d2*(xx(m))^2); end
plot(xx,yy,'k–', x,y,'ko','LineWidth',2.0)

The recovered function shown in Fig. 4.4, is exactly the same function $f(x) = \frac{1}{1+10x^2}$ that was used to generate the data. That is one finds $b_1 = 1$ & $b_2 = 10$. The above example again shows that if data points belong to a function which is a member of a chosen family of functions, then the least square fit will find a function which goes through all the data points. A question motivated by this least square result is what happens if data deviate a little from the family of functions being sampled. For example, in the case when data contains measurement errors. To demonstrate this, let us take data points (x_k, y_k) shown in Table 4.2, but changed its y-values randomly from

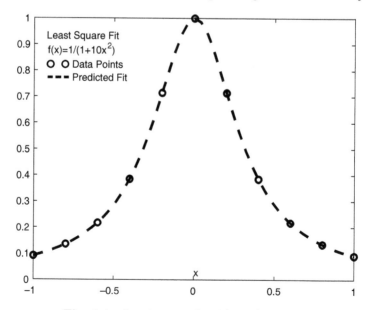

Fig. 4.4. Least square fit with no data error.

Table 4.3. Same as Table 4.2, except for small variations in data points (x_k, y_k).

n	x_n	y_n	$\tilde{y}_k$	x_n^2	x_n^4	$x_n^2 \tilde{y}_n$
1	-1	0.09	11.11	1	1	11.11
2	-0.8	0.14	7.14	0.64	0.4096	4.57
3	-0.6	0.24	4.17	0.36	0.1296	1.50
4	-0.4	0.35	2.86	0.16	0.0256	0.46
5	-0.2	0.68	1.47	0.04	0.0016	0.06
6	0	1.07	0.93	0	0	0.00
7	0.2	0.79	1.27	0.04	0.0016	0.05
8	0.4	0.36	2.78	0.16	0.0256	0.44
9	0.6	0.23	4.35	0.36	0.1296	1.57
10	0.8	0.15	6.67	0.64	0.4096	4.27
11	1	0.08	12.50	1	1	12.50
Sum	0	4.18	55.24	4.40	3.13	36.53

their exact values, by at most 10%. The quantities needed to find the least square fit to this modified data are presented in Table 4.3, and its graph shown in Fig. 4.5.

Table 4.3 shows the actual data with at most 10% random error added. The best fit to this modified data set is shown in Fig. 4.5.

Figure 4.5 shows the least square fit to data points containing some measurement errors. The function that fits this modified data in least square sense has the form

$$\tilde{f}(x) = \frac{1}{0.8176 + 10.5107x^2}$$

which is not the same as the initial function $f(x) = \frac{1}{1+10x^2}$. As expected, predicted function does not go through all the data points that do not have measurement error but is still a member of the specified family of functions of the form: $f(x) = \frac{1}{b_1 + b_2 x^2}$. The modified b_1, b_2 due to existence of 10% measurement error are different from actual constants b_1, b_2, by -18.24% and 5.11%, respectively.

Above examples show that the least square procedure is mainly used when a theory with some unknown parameters is developed, but the theory is in need of specifying some unknown parameters from the actual measured data, in order to make the theory complete so it can be used in applications. A good example is finding gravitational constant "g" from motion of a free falling object.

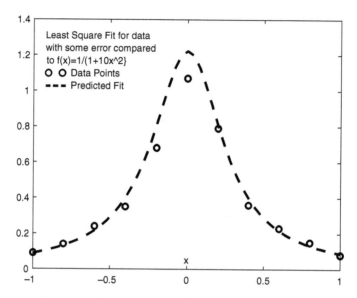

Fig. 4.5.   Least square fit for data with some errors.

The above examples demonstrate that to find the unknown parameters in a theory as accurately as possible, data measurement related to the theory needs to be found accurately.

4.5   Fourier Series and Trigonometric Polynomials

Another interpolation/approximation of data is achieved via Fourier series. The procedure is developed for periodic functions, that is $f(x + \tau) = f(x)$, and τ is the function's period. The method makes extensive use of orthogonality properties of sin and cos functions. That is,

$$\int_{-\pi}^{\pi} \cos nx \cos mx\, dx = \pi \delta_{nm}$$

$$\int_{-\pi}^{\pi} \sin nx \sin mx\, dx = \pi \delta_{nm}$$

$$\int_{-\pi}^{\pi} \cos nx \sin mx\, dx = 0$$

where the Kronecker

$$\delta_{nm} = \begin{cases} 1 & \text{if } m = n \\ 0 & \text{if } m \neq n \end{cases}$$

Of course one can also apply this procedure to a function defined on a finite interval, i.e. $[a, b]$ and think of the interval as one period of the given function.

Definition. Let $f(x)$ defined over $[-\pi, \pi]$ be a continuous function except at a finite number of points in $[-\pi, \pi]$ and $f(x)$ is a periodic function with a period of 2π, then its Fourier series $S(x)$ over the interval $[-\pi, \pi]$ is defined as

$$S(x) = \frac{a_0}{2} + \sum_{j=1}^{\infty} (a_j \cos(jx) + b_j \sin(jx))$$

with

$$a_j = \frac{1}{\pi} \int_{-\pi}^{\pi} f(x) \cos(jx) dx$$

$$b_j = \frac{1}{\pi} \int_{-\pi}^{\pi} f(x) \sin(jx) dx$$

$j = 0, 1, 2, \ldots$ Above relations are based on orthogonality properties of sin and cos.

Fourier expansion theorem

If $f(x)$ is a piecewise continuous function which is 2π periodic on the closed interval $[-\pi, \pi]$ and both left and right derivatives at each $x \in [-\pi, \pi]$ exist, then for each $x \in [-\pi, \pi]$ the Fourier series $S(x)$ of $f(x)$ converges to the value $\frac{f(x^-) + f(x^+)}{2}$.

Definition. $f(x)$ defined over interval $[c, d]$ is called piecewise continuous, if it is continuous over $[c, d]$ except for a finite number of points $x \in [c, d]$.

Example. Consider the function $f(x)$ defined over $[-\pi, \pi]$, when

$$f(x) = \begin{cases} 1 & \text{for } x \in [a, b] \ \& \ -\pi < a < b < \pi \\ 0 & \text{otherwise} \end{cases}$$

$$a_j = \frac{1}{\pi} \int_{-\pi}^{\pi} f(x) \cos(jx) dx = \frac{1}{\pi} \int_{a}^{b} \cos(jx) dx$$

$$b_j = \frac{1}{\pi} \int_{-\pi}^{\pi} f(x) \sin(jx) dx = \frac{1}{\pi} \int_a^b \sin(jx) dx$$

$$\text{for } j = 0, \ a_0 = \frac{(b-a)}{\pi}$$

$$\text{and for } j \neq 0, \ a_j = \frac{1}{\pi j}(\sin(jb) - \sin(ja))$$

$$b_j = -\frac{1}{\pi j}(\cos(jb) - \cos(ja))$$

Trigonometric polynomial of order M

Since numerically we can only add finite number of terms appearing in the Fourier series $S(x)$, following approximation is defined:

$$P_M = \frac{a_0}{2} + \sum_{j=1}^{M}(a_j \cos(jx) + b_j \sin(jx))$$

This approximation of $f(x)$ can be easily computed for the example given, but for general $f(x)$ one needs to carry out numerical integration, which will be covered in chapter V. So for now let's see how the defined Trigonometric polynomials perform for above example when using different values of M, i.e. for $M = 5$ and $M = 50$, when $b = -a = \frac{\pi}{2}$

$$P_5 = \frac{1}{2} + \frac{1}{\pi} \sum_{j=1}^{5} \frac{2}{j} \sin\left(\frac{j\pi}{2}\right) \cos(jx)$$

Figure 4.6 shows the result of Fourier expansion for two trigonometric polynomials, $M = 5$ and $M = 50$. It shows recovery of the function $f(x)$ becomes better as M increased. But P_5 nicely shows the relation $S(a) = \frac{f(a^-)+f(a^+)}{2}$, since $f(x)$ has a jump at $x = a = -\frac{\pi}{2}$. P_{50} nicely shows approximation has less error, but it also has oscillation near discontinuity of $f(x)$, which is called the Gibbs phenomenon.

In order to estimate order of trigonometric polynomial needed to approximate $f(x)$, within a given error tolerance, one defines the L_2

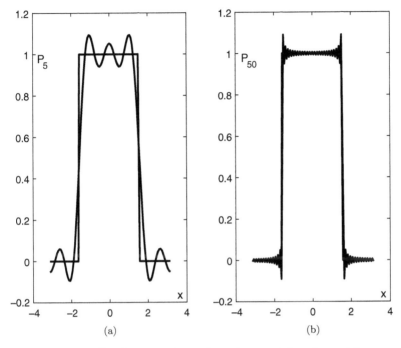

Fig. 4.6. Fourier expansion using trigonometric polynomials.

norm of the difference

$$E_M = \left[\int_{-\pi}^{\pi} (f(x) - P_M(x))^2 \, dx \right]^{\frac{1}{2}}$$

$$= \left[\int_{-\pi}^{\pi} \left(f(x) - \sum_{n=0}^{M} \tilde{a}_n \cos nx + \tilde{b}_n \sin nx \right)^2 dx \right]^{\frac{1}{2}}$$

where $\tilde{a}_0 = \frac{a_0}{2}$, $\tilde{b}_0 = 0$, $\tilde{a}_n = a_n$, $\tilde{b}_n = b_n$, for $n = 1, 2, \ldots, M$. Thus,

$$E_M^2 = \int_{-\pi}^{\pi} \left(f(x) - \sum_{k=0}^{M} \tilde{a}_k \cos kx + \tilde{b}_k \sin kx \right)^2 dx$$

$$= \int_{-\pi}^{\pi} f^2(x) dx - 2 \int_{-\pi}^{\pi} f(x) \left(\sum_{k=0}^{M} \tilde{a}_n \cos nx + \tilde{b}_n \sin nx \right) dx$$

$$+ \int_{-\pi}^{\pi} \left(\sum_{k=0}^{M} \tilde{a}_k \cos kx + \tilde{b}_k \sin kx \right)$$

$$\times \left(\sum_{n=0}^{M} \tilde{a}_n \cos nx + \tilde{b}_n \sin nx \right) dx$$

Making use of orthogonality properties of sin and cos, one finds

$$E_M^2 = \int_{-\pi}^{\pi} f^2(x)dx - \pi a_0^2 - 2\pi \sum_{n=1}^{M} [a_n^2 + b_n^2]$$

$$+ \int_{-\pi}^{\pi} \left(\frac{a_0^2}{4} + \sum_{n=1}^{M} a_n^2 \cos^2 nx + b_n \sin^2 nx \right) dx$$

$$E_M = \left(\int_{-\pi}^{\pi} f^2(x)dx - \frac{\pi a_0^2}{2} - \pi \sum_{n=1}^{M} (a_n^2 + b_n^2) \right)^{\frac{1}{2}}$$

From the above relation order of the trigonometric polynomial can be deduced for a given error tolerance E_M. The above relation when M tends to infinity and E_M tends to zero, is called Parseval's identity. That is,

$$\frac{1}{\pi} \int_{-\pi}^{\pi} f^2(x)dx = \frac{a_0^2}{2} + \sum_{n=1}^{\infty} (a_n^2 + b_n^2)$$

Above derivation shows that equality of $f(x)$ in terms of its Fourier series representation is in the sense of L_2 norm.

4.6 Discrete Fourier Series

There are cases of interest when $f(x)$ is only known for discrete values of x, or when Fourier series can be computed much faster, called fast Fourier transform, which necessitates sampling $f(x)$ at specified discrete points. For our example, let us take $x_j = -\pi + \frac{2j\pi}{N}$, for $j = 0, 1, 2, \ldots, (N-1)$. Let $f(x)$ be 2π periodic, and $N > 2M$, with M being the order of the trigonometric polynomial P_M. Discrete

Fourier series of $f(x)$ is then defined as

$$P_M(x) = \frac{a_0}{2} + \sum_{j=1}^{M} a_j \cos(jx) + b_j \sin(jx)$$

with

$$a_j = \frac{2}{N} \sum_{k=0}^{N-1} f(x_k) \cos(jx_k), \quad j = 0, 1, 2, \ldots M$$

$$b_j = \frac{2}{N} \sum_{k=0}^{N-1} f(x_k) \sin(jx_k), \quad j = 0, 1, 2, \ldots M$$

In other words, for discrete Fourier series of $f(x)$, one does not need to carry out any integration to find coefficients a_j & b_j but one needs to choose sampling rate N to be more than twice as large as the degree of trigonometric polynomial $P_M(x)$.

As an example, consider Fig. 4.7 that shows result of approximation if we choose $M = 10$ when sampling frequency $N = 22 > 2(10) = 2M$. Another example for the same function $f(x)$ the trigonometric polynomial P_{10} is calculated when $N = 10 < 2(10) = 2M$.

In Fig. 4.7, the function $f(x)$ is same as in Fig. 4.6, where

$$f(x) = \begin{cases} 1 & a \leq x \leq b \text{ where } -\pi < a < b < \pi \\ 0 & \text{otherwise} \end{cases}$$

The main difference in the above two figures is that we are finding discrete Fourier series of $f(x)$ by sampling, $f(x)$ at $x_j = -\pi + \frac{2j\pi}{N}, j = 0, 1, 2, \ldots, (N-1)$, but one with $N = 22$, and the other $N = 10$ for finding P_{10}. As it can be seen from Fig. 4.7 performance is satisfactory, if $N > 2M$, but the approximation is not satisfactory when $N \leq 2M$.

For the last example of this chapter let us see how discrete Fourier series performs when applied to our previous example $f(x) = \frac{1}{1+10x^2}$ with same sampling points as before, $x_n = -1 + \frac{2n}{N}$, with $n = 0, 1, 2 \ldots N = 10$. Using $M = 4$, will satisfy $N > 2M$ condition. Since

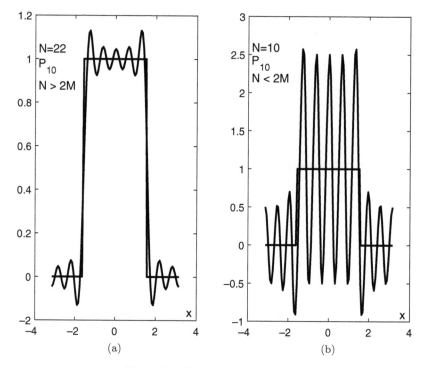

Fig. 4.7. Discrete fourier series.

$f(x)$ is defined over $[-1,1]$ we also need to make scale transformation. To achieve this, define $t_n = \pi x_n$, and compare the results with other procedures previously presented. We will use the same data as used previously for $y_n = \frac{1}{1+10x_n^2}$.

Comparing the result shown in Fig. 4.8 to Figs. 4.2 and 4.4, all approximating the same function $f(x) = 1/(1+10x^2)$, one notes the discrete Fourier series does as well as other procedures, even though the sampling rate was small, thanks to the smoothness of $f(x)$.

MATLAB code for generating Fig. 4.8

```
% imputing initial information N=10; M=4; N1=N+1; M1=M+1;
NT=100; dx=2/N; dtx=2/NT;
% defining nodal points
for n=1:N1
x(n)=-1+(n-1)*dx; y(n)=1/(1+10*(x(n))^2); end
% defining coefficients
```

```
for j=1:M1
aa=0; bb=0;
for n=1:N
aa=aa+y(n)*cos((j-1)*pi*x(n)); bb=bb+y(n)*sin((j-1)*pi*(n)); end
a(j)=(2/N)*aa; b(j)=(2/N)*bb; end
% finding the transform
for k=1:NT
xx(k)=-1+(k-1)*dtx; TK=a(1)/2; for j=2:M1
TK= TK +a(j)*cos((j-1)*pi*xx(k))+b(j)*sin((j-1)*pi*xx(k)); end
TM(k)=TK; end
plot(xx,TM, 'k-',x,y,'kO','linewidth',1.5)
text(-0.85,.95, 'N=10')
text(-0.85,0.87, 'T₄')
```

Comparing Figs. 4.4 and 4.8 for the same sampled data shows that results for the two procedures are very similar, although Fourier series procedure did not directly enforce the requirement that the found approximate function needs to pass through all the sampled data points.

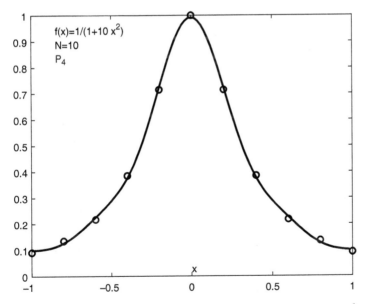

Fig. 4.8. Discrete Fourier series for sampled data from $f(x) = \frac{1}{1+10x^2}$.

Exercises

4.1. $f(x) = \frac{1}{1+10x^2}$ is sampled at points $x_n \in [-1, 1]$, $n = 0, 1, 2, 3, 4$. x_n are equally spaced and $x_0 = -1$ and $x_4 = 1$. Find Lagrange Interpolating Polynomials $P_4(x)$ for $f(x)$ based on data for the sampled nodes. Graph $E_4(x) = f(x) - P_4(x)$ and compare it with the given estimate of $E_4(x)$.

4.2. Prove that $R_N(x) \equiv f(x) - P_N(x) = \frac{\phi(x)}{(N+1)!} f^{(N+1)}(\xi)$, where $\phi(x) = (x - x_0)(x - x_1) \cdots (x - x_N)$ and $\xi = \xi(x) \in (a, b)$.

Hints: (i) define $\psi(x) = R_N(x) - \frac{\phi(x)}{\phi(\tilde{x})} R_N(\tilde{x})$, where the arbitrary constant $\tilde{x} \neq x_n$ with $\tilde{x} \in [a, b]$ & $n = 0, 1, \ldots, N$. (ii) verify $\psi(x_n) = 0$, $\psi(\tilde{x}) = 0$ and $\psi(x) \in C^{N+1}[a, b]$. (iii) show that $\psi(x)$ has $N+2$ distinct roots, and $\psi'(x)$ has $N+1$ distinct roots. (iv) by induction show that $\psi^{(N+1)}$ has a zero labeled as $\xi \in (a, b)$ & prove $\psi^{(N+1)}(\xi) = R_N^{(N+1)}(\xi) - \frac{(N+1)!}{\phi(\tilde{x})} R_N(\tilde{x}) = 0$. (v) complete the proof that $R_N(x) = \frac{\phi(x)}{(N+1)!} f^{(N+1)}(\xi)$.

4.3. Same as Problem 4.1, except sample $f(x)$, using Chebyshev nodes. Graph $E_4(x)$, when Chebyshev nodes are used.

4.4. Interpolate the following data: $(x_0, 0.1), (x_1, 0.2), (x_2, 1), (x_3, 0.2), (x_4, 0.1)$, using Chebyshev procedure with $x_0 = 0$, $x_4 = 8$ and x_1, x_2, x_3 are the other Chebyshev nodes for $x \in [0, 8]$.

4.5. Use the least square procedure to find the best fit parameters α and β appearing in $f(x) = \frac{\alpha}{1+\beta x^3}$. Use data points $\{(x_n, e^{-x_n^3})\}$ for $x_n = -1 + 0.5n$, when $n = 0, 1, 2, 3, 4$.

4.6. Use the least square procedure to find the best fit to the data points: $\{(-3, 0.003), (-1.5, 0.05), (0, 1), (1.5, 0.05), (3, 0.003)\}$, by the function $f(x) = \alpha e^{\beta x^2}$, where α and β are parameters to be found by the least square procedure.

4.7. Use the least square procedure to find the best fit to the data points: $\{(1, 2.0), (1.5, 4.1), (2, 7.0), (2.5, 10.5), (3, 14.9)\}$ by the function $f(x) = \alpha x^\beta$, where α and β are parameters to be found by the least square procedure.

4.8. Use the least square procedure to find the best fit to the data points:
$(x, y, z) = \{(-3, 3, 6), (-1, -1, 0), (0, -2, -3), (1, 2, 4), (2, 3, 2)\}$

by the plane $z = \alpha x + \beta y + \gamma$, where α, β and γ are the parameters to be found by the least square procedure.

4.9. Same as Problem 4.1 except use Natural Cubic Spline, $S(x)$, to interpolate $f(x)$ using the given nodes. Also, graph $E(x) = f(x) - S(x)$.

4.10. The function $f(x) = -|x| + 2$ is sampled at points $x_n \in [-1, 1]$, $n = 0, 1, 2, 3, 4, 5$. x_n are equally spaced with $x_0 = -1$ and $x_5 = 1$. Use Clamped Cubic Spline $S(x)$ to interpolate $f(x)$ for sampled nodes x_n, with the assumption that $S'(-1) = 1$ & $S'(1) = -1$.

4.11. Prove that:

   (a) $\int_{-\pi}^{\pi} \cos nx \, \cos mx \, dx = \pi \delta_{nm}$

   (b) $\int_{-\pi}^{\pi} \sin nx \, \sin mx \, dx = \pi \delta_{nm}$

   (c) $\int_{-\pi}^{\pi} \cos nx \, \sin mx \, dx = 0$,

where $\delta_{nm} = \begin{cases} 1 & \text{if } m = n \\ 0 & \text{if } m \neq n \end{cases}$

4.12. Let 2π periodic $f(x) = -|x| + \pi$ with one period being over interval $[-\pi, \pi]$.

   (a) Find its Fourier trigonometric polynomial representation P_{30}.

   (b) Find M, the smallest order of trigonometric polynomial P_M approximating $f(x)$ with its L_2 error $E_M \leq 0.05$.

4.13. Let $f(x) = e^{-|x|}$ be a periodic function with one period being over interval $[-\pi, \pi]$.

   (a) Write its discrete Fourier series formula P_{10} for above $f(x)$.

   (b) Use equally spaced sampling of the function with lowest total sampling number N that results in an acceptable discrete Fourier series approximation of $f(x)$.

   (c) Numerically evaluate your discrete Fourier series P_{10} for $f(x)$, using sampling number N, you selected in part (b).

   (d) Graph $f(x)$ and its corresponding approximation P_{10}.

4.14. Let $f(x) = e^{-x^2}$, be define as a 2π periodic function sampled at points $x_n \in [-\pi, \pi]$, for $n = 0, 1, 2, \ldots, 12$, where x_n are equally spaced with $x_0 = -\pi$ and $x_{12} = \pi$. Find the discrete Fourier series P_5 for $f(x)$, and graph both $P_5(x)$ and $f(x)$.

4.15. Let a saw tooth periodic $f(x) = -|x| + 1$ with one period defined over interval $[-1, 1]$ be sampled at $x_n = -1 + nh$, $n = 0, 1, \ldots, 30$, & $h = \frac{2}{30}$. Find the highest value of M such that trigonometric polynomial P_M will find a satisfactory discrete Fourier series for $f(x)$. Graph your P_M and $f(x)$.

Chapter 5

Differentiation and Integration

Another area that plays a crucial role in countless areas of scientific endeavors is differentiation and integration. Starting with the well-known definition of derivative

$$f'(x) = \lim_{h \to 0} \frac{f(x+h) - f(x)}{h}$$

we note as it stands this definition is not numerically practical for calculating derivatives, since one needs to take the limit as h tends to zero. Thus, in the next section we study how to implement the concept of differentiation numerically.

5.1 Numerical Differentiation

To differentiate a function numerically use is made of Taylor expansion. That is, if $f(x)$ is suitably differentiable, then according to Taylor expansion

$$f(x_o + h) = f(x_o) + f'(x_o)h + \frac{f''(x_o)}{2}h^2 + \frac{f'''(x_o)}{6}h^3 + \cdots$$

Thus,

$$\frac{f(x_o + h) - f(x_o)}{h} = f'(x_o) + \frac{f''(x_o)}{2}h + O(h^2) = f'(x_o) + O(h)$$

or

$$f'(x_o) = \frac{f(x_o + h) - f(x_o)}{h} + O(h)$$

The above expression gives us an approximate value of the derivative and an estimate of its associated error. The above definition of numerical derivative is called Forward Difference approximation of the derivative.

Another approximation of the derivative that does not require h to be excessively small for a desired accuracy, is to again apply Taylor expansion, but note that

$$f(x_o \pm h) = f(x_o) \pm f'(x_o)h + \frac{1}{2}h^2 f''(x_o)h^2 + O(h^3)$$

Subtracting above two equations from each other leads to

$$\frac{f(x_o + h) - f(x_o - h)}{2h} = f'(x_o) + O(h^2)$$

or

$$f'(x_o) = \frac{f(x_o + h) - f(x_o - h)}{2h} + O(h^2)$$

This approximation is called Central Difference approximation of the derivative. It can be seen from the above presentation that the truncation error for forward difference approximation is $O(h) \approx \frac{f''(c)}{2}h$ and for central difference approximation is $O(h^2) \approx \frac{f'''(c)}{6}h^2$. Higher-order Taylor expansion can also be used to find $f'(x_o)$ more accurately, but use of forward difference or central difference approximations are most common. A practical way to find a more accurate representation of numerical derivative is to use the following method.

5.2 Richardson Extrapolation Method

The Richardson procedure applies to any approximation of a quantity $\tilde{R}$ by $R(h)$ for small values of h, if they satisfy following relation:

$$R(h) = \tilde{R} + Ch^n + O(h^{n+1})$$

with C being a constant. Approximating $\tilde{R}$ by a different step size $\frac{h}{\alpha}$ where α is a constant greater than one, leads to

$$R\left(\frac{h}{\alpha}\right) = \tilde{R} + C\left(\frac{h}{\alpha}\right)^n + O\left(\left(\frac{h}{\alpha}\right)^{n+1}\right),$$

Multiplying above equation by α^n and subtraction of original equation from it, leads to

$$R(h, \alpha) \equiv \frac{\alpha^n R(\frac{h}{\alpha}) - R(h)}{\alpha^n - 1} = \tilde{R} + O(h^{n+1})$$

or

$$\tilde{R} = R(h, \alpha) + O(h^{n+1})$$

Above equation shows the defined $R(h, \alpha) = \frac{\alpha^n R(\frac{h}{\alpha}) - R(h)}{\alpha^n - 1}$, called Richardson extrapolation, approximates $\tilde{R}$ more accurately than the original approximation by $R(h)$, when h is small.

To show its application let us consider forward difference approximation as an example. Using Richardson extrapolation notation, we have

$$R(h) = \frac{f(x_o + h) - f(x_o)}{h}, \tilde{R} = f'(x_o), C = \frac{f''(x_o)}{2}, \& \ n = 1.$$

Taking $\alpha = 2$ Richardson extrapolation requires finding

$$R(h, 2) = \frac{2R(\frac{h}{2}) - R(h)}{2 - 1} = f'(x) + O(h^2),$$

for estimating f'. Above results shows by finding forward difference for two different step sizes to compute $R(h, 2)$ makes forward difference approximation accuracy to improve from $O(h)$ to $O(h^2)$.

Numerical Examples. For $f(x) = e^{-x^2} + x^3$ let us compare (i) forward difference, (ii) central difference derivatives and (iii) Richardson extrapolation to find $f'(x)$ at $x = 0$, noting Exact $f'(0) = 0$ and taking $h = 0.2$ we have

(1) Forward difference:

$$f'(0) \approx \frac{f(.2) - f(0)}{0.2} = R(.2) = -0.1561 = O(h) \quad \text{for } h = 0.2$$

$$f'(0) \approx \frac{f(.1) - f(0)}{0.1} = R(.1) = -0.0895 = O(h) \quad \text{for } h = 0.1$$

(2) Central difference:

$$f'(0) \approx \frac{f(.2) - f(-0.2)}{0.4} = 0.0400 \approx h^2 = O(h^2) \quad \text{for } h = 0.2$$

(3) Richardson extrapolation:

$$f'(0) = 2R(.1) - R(.2) \approx -0.0230 = O(h^2)$$
$$\text{for } h = 0.2 \ \& \ \alpha = 2$$

As it can be seen from the above, for $f(x) = e^{-x^2} + x^3$, errors in calculating $f'(0)$ using above procedures are all in agreement with our expectations, with Richardson extrapolation performing a little better than central difference approximation, if we do not take into account additional computation needed to find Richardson extrapolation result.

Higher derivatives are also found numerically by making use of Taylor expansion.

For example,

$$f(x_o \pm h) = f(x_o) \pm f'(x_o)h + f''(x_o)\frac{h^2}{2} \pm f'''(x_o)\frac{h^3}{6} + O(h^4)$$

add the above two equations:

$$f(x_o + h) + f(x_o - h) = 2f(x_o) + f''(x_o)h^2 + O(h^4)$$

Thus,

$$f''(x_o) = \frac{f(x_o + h) - 2f(x_o) + f(x_o - h)}{h^2} + O(h^2)$$

Since in practice data for $f(x_n)$ are only available for a set $S = \{x_n\}$, where $n = 1, 2, 3, \ldots$, the above procedures will not be numerically feasible without further approximation if $(x_o \pm h) \notin S$. In such a

situations a useful procedure is to first find a function that interpolates the data, and then analytically take derivative of the found
function to estimate $f'(x_0)$. As an example, let us take the example
shown in Fig. 4.7, and assume data $(x_n, f(x_n))$ are only available for
$f(x_n) = 1/(1 + 10x_n^2)$ for $x_n = -1 + 0.2n$, for $n = 0, 1, 2, \ldots, 9, 10$.
This information enables us to first find the trigonometric polynomial

$$T_4(x) = \frac{a_0}{2} + \sum_{j=1}^{4}(a_j \cos(jx) + b_j \sin(jx))$$

To estimate $f'(x)$ at $x = -0.525 \notin S$, since the point $x = -0.525$ is
not one of the given data points x_n, above presented finite difference
methods cannot be used to find $f'(-0.525)$, but $T_4(x)$ can be used
to find $f'(-0.525)$. Since coefficients a_j and b_j have been already
calculated from the given data, and

$$f'(x) \approx T_4'(x) = \sum_{j=1}^{4} j(-a_j \sin(jx) + b_j \cos(jx))$$

Substitution of $x = -0.525$ in the above equation leads to
approximate value of $f' \approx T_4' = 0.6495$. We know the exact value
of $f' = 0.7442$. Thus, the difference between f' and T_4' is $0.0947 \approx$
$0.2^{1.46}$, consistent with the size of $h = 0.2$, especially in view of small
number of data points, and low order trigonometric polynomials used
to approximate $f(x)$.

5.3   Numerical Integration

The basic idea on how to find a definite integral

$$I[f] = \int_a^b f(x)dx$$

numerically is to approximate $f(x)$ by $\tilde{f}(x)$ whose integral is
analytically known. $\tilde{f}(x)$ is usually defined over N subdivisions of
the interval $[a, b]$, i.e. $[x_n, x_{n+1}]$, where $a = x_0 < x_1 < x_2 < \cdots <$
$x_n \cdots < x_N = b$. The most familiar example of integration procedure
introduced in calculus is approximating $f(x)$ by

$$\tilde{f}(x) = f\left(\frac{x_{n+1} + x_n}{2}\right), \quad x \in [x_n, x_{n+1}]$$

which is the mid-value of $f(x)$ over each subinterval $[x_n, x_{n+1}]$, and an approximate integral of $f(x)$ is found by adding areas of all rectangles formed by this procedure.

A systematic procedure for such an estimate of integral of f over $[a, b]$, is called the quadrature of f and denoted as

$$Q[f] = \sum_{n=0}^{N} w_n f(x_n)$$

The constants w_n is called the weights of the quadrature, with nodes x_n defined for values of $n = 0, 1, \ldots, N$. Usually, nodes are chosen to be equally spaced. $x_n = a + nh$ & $h = \frac{b-a}{N}$. The truncation error generated by approximating integral of $f(x)$ by its quadrature Q is denoted as

$$E[f] = I[f] - Q[f] = \int_a^b f(x)dx - \sum_{n=0}^{N} w_n f(x_n)$$

The truncation error $E[f]$ can be estimated by finding the degree of precision of the quadrature. The degree of precision is defined to be the maximum integer m such that

$$E[x^l] = 0, \quad \text{for } l = 0, 1, 2, \ldots, m, \quad \text{but } E[x^{m+1}] \neq 0$$

Theorem. *If $Q[f] \approx \int_{x_n}^{x_{n+1}} f(x)dx$ has a precision degree m, and $f(x)$ is $m + 1$ times differentiable, then its truncation error $E[f] = O(h^{m+2})$, where $x_n = a + nh$ & $h = x_{n+1} - x_n$.*

Proof. It follows from Taylor's theorem that for $x_n \leq x < x_{n+1}$

$$f(x) = \tilde{f}(x) + \tilde{R}(x) \quad \text{where } \tilde{f}(x) = \sum_{k=0}^{m} \frac{f^{(k)}(x_n)}{k!}(x - x_n)^k$$

and

$$\tilde{R}(x) = \frac{f^{(m+1)}(\xi_n)}{(m+1)!}(x - x_n)^{m+1} \quad \text{with } x_n \leq \xi_n \leq x_{n+1}$$

If the precision is m, then $\int_{x_n}^{x_{n+1}} \tilde{f}(x)dx = Q[\tilde{f}]$, since $E[x^l] = \int_{x_n}^{x_{n+1}} x^l dx - Q[x^l] = 0$, for $l = 0, 1, 2, \ldots, m$. Thus,

$$I[f] = \int_{x_n}^{x_{n+1}} f(x)dx = \int_{x_n}^{x_{n+1}} \tilde{f}(x)dx + \int_{x_n}^{x_{n+1}} \tilde{R}(x)dx$$

$$= Q[f] + O(h^{m+2})$$

Before giving examples of precision degree of a quadrature, let us note two types of quadratures are commonly used. They are open and closed quadratures. For an open quadrature values of the function at nodal end points do not appear in the definition of quadrature, but for a closed quadrature the end point nodes are used in the definition of $Q[f]$. The followings are examples of open and closed quadratures defined for integral of $f(x)$ over the interval $[x_n, x_{n+1}]$ of length $h = x_{n+1} - x_n$.

Rectangular quadrature for interval $[x_n, x_{n+1}]$ to be an open quadrature, by definition, it should not contain values of $f(x)$ at the end points x_n and x_{n+1}. To achieve this one may define the rectangular quadrature as

$$Q_R(f) = hf\left(\frac{x_{n+1} + x_n}{2}\right)$$

where only the mid-point value of the interval $[x_n, x_{n+1}]$ is used in defining the rectangular quadrature of $f(x)$, over the subinterval $[x_n, x_{n+1}]$.

Trapezoidal quadrature is a closed quadrature by its definition

$$Q_T(f) = h\frac{f(x_n) + f(x_{n+1})}{2} = w_n f(x_n) + w_{n+1} f(x_{n+1}), \quad w_n = \frac{h}{2}$$

Note values of $f(x)$ at nodal points x_n & x_{n+1}, which are the end points of subinterval $[x_n, x_{n+1}]$ are used in defining $Q_T(f)$.

Let us now find precision of above defined quadratures.

1. Rectangular quadratures: $\int_{x_n}^{x_{n+1}} f(x)dx = Q_R[f] + E_R[f]$

For $l = 0$, $f(x) = x^l = x^0 = 1$, $I[f] = \int_{x_n}^{x_{n+1}} x^0 dx = x_{n+1} - x_n = h$

$Q_R[f] = hf\left(\dfrac{x_n + x_{n+1}}{2}\right) = h = I[f] \Rightarrow E_R[f] = 0$

For $l = 1$, $f(x) = x^l = x$, $\int_{x_n}^{x_{n+1}} x dx = \dfrac{x_{n+1}^2 - x_n^2}{2}$

$= (x_{n+1} - x_n)\dfrac{x_{n+1} + x_n}{2} = hf\left(\dfrac{x_{n+1} + x_n}{2}\right) = Q_R[x] \Rightarrow E_R[f] = 0$

For $l = 2$, $f(x) = x^2$, $\int_{x_n}^{x_{n+1}} x^2 dx = \dfrac{x_{n+1}^3 - x_n^3}{3}$

$= \dfrac{(x_{n+1}^2 + x_n x_{n+1} + x_n^2)(x_{n+1} - x_n)}{3} = \dfrac{h}{3}(x_{n+1}^2 + x_n x_{n+1} + x_n^2)$

$\neq Q_R[f] = hf\left(\dfrac{x_{n+1} + x_n}{2}\right) = h\left(\dfrac{x_{n+1} + x_n}{2}\right)^2$

$= \dfrac{h}{4}(x_{n+1}^2 + 2x_n x_{n+1} + x_n^2) \Rightarrow E_T[f] \neq 0$

Thus, by definition of quadrature precision, it follows that degree of precision for a rectangular quadrature $Q_R[f]$ is one. In other words, $m = 1$.

Next, let us compute degree of precision for a trapezoidal quadrature.

2. Trapezoidal quadrature: $\int_{x_n}^{x_{n+1}} f(x)dx = Q_T[f] + E_T[f]$

For $l = 0$, $f(x) = x^l = x^0 = 1$, $I[f] = \int_{x_n}^{x_{n+1}} x^0 dx = x_{n+1} - x_n = h$

$Q_T[f] = \dfrac{h}{2}f(x_n) + \dfrac{h}{2}f(x_{n+1}) = \dfrac{h}{2} + \dfrac{h}{2} = h = I[f] \Rightarrow E_T[f] = 0$

For $l = 1$, $f(x) = x$, $I[f] = \int_{x_n}^{x_{n+1}} x \, dx = \dfrac{x_{n+1}^2 - x_n^2}{2}$

$= (x_{n+1} - x_n) \dfrac{x_{n+1} + x_n}{2} = \dfrac{h}{2} f(x_{n+1}) + \dfrac{h}{2} f(x_n) = Q_T[x] \Rightarrow E_T[f] = 0$

For $l = 2$, $f(x) = x^2$, $I[f] = \int_{x_n}^{x_{n+1}} x^2 \, dx$

$= \dfrac{x_{n+1}^3 - x_n^3}{3} = \dfrac{(x_{n+1}^2 + x_n x_{n+1} + x_n^2)(x_{n+1} - x_n)}{3}$

$= \dfrac{h}{3}(x_{n+1}^2 + x_n x_{n+1} + x_n^2) \neq Q_T[f] = \dfrac{h}{2} f(x_n) + \dfrac{h}{2} f(x_{n+1})$

$= \dfrac{h}{2}(x_{n+1}^2 + x_n^2) \Rightarrow E_T[f] \neq 0$

Hence, the degree of precision for trapezoid quadrature is also $m = 1$.

Above examples motivates studying quadratures with higher degree of precision, realizing these well-know quadratures we presented made a linear approximation of a function over successive nodal points. For example, by joining values of the function at every successive end points of a panel by a straight line gave us the trapezoidal quadrature rule with precision $m = 1$. Thus, if we want to define a quadrature with more precision, an N point representation of the function via its Lagrange polynomials representation could guide us on how to develop such a quadrature with higher degree of precision. To achieve this let

$$f(x) = P_N(x) + R_N(x), \text{ where } P_N(x) = \sum_{k=0}^{N} f(x_k) L_{N,k}(x)$$

$$L_{N,k} = \Pi_{j=0, j \neq k}^{N} \dfrac{(x - x_j)}{(x_k - x_j)} \ \&$$

$$R_N(x) = \dfrac{[(x - x_0) \cdots (x - x_N)] f^{(N+1)}(\xi)}{(N + 1)!}$$

Using $P_N(x)$ with different N values allows us to define quadratures with different degrees of precision. For example, if we take $N = 1$, we find

$$P_1(x) = \sum_{k=0}^{1} f(x_k)L_{1,k}(x) = f(x_0)L_{1,0} + f(x_1)L_{1,1}(x)$$

$$= f(x_0)\frac{(x - x_1)}{(x_0 - x_1)} + f(x_1)\frac{(x - x_0)}{(x_1 - x_0)}$$

$$= \frac{f(x_1) - f(x_0)}{x_1 - x_0}(x - x_0) + f(x_0)$$

and $R_1(x) = \dfrac{(x - x_0)(x - x_1)f''(\xi)}{2!}$

$$= \frac{(x^2 - (x_1 + x_0)x + x_1x_0)f''(\xi)}{2}$$

Therefore,

$$\int_{x_0}^{x_1} f(x)dx = \int_{x_0}^{x_1} P_1(x)dx + \int_{a}^{b} R_1(x)dx = Q[f] + E[f]$$

$$Q[f] = \int_{x_0}^{x_1} P_1(x)dx$$

$$= \int_{x_1}^{x_2} \left[\frac{f(x_1) - f(x_0)}{x_1 - x_0}(x - x_0) + f(x_0) \right] dx$$

$$= \frac{f(x_1) - f(x_0)}{2(x_1 - x_0)}(x_1 - x_0)^2 + f(x_0)(x_1 - x_0)$$

$$= \frac{f(x_1) + f(x_0)}{2}(x_1 - x_0)$$

Hence, the quadrature defined via Lagrange interpolation for $N = 1$ is the well-known trapezoidal quadrature

$$Q_T[f] = \sum w_i f(x_i) = w_0 f(x_0) + w_1 f(x_1)$$

where $w_0 = w_1 = \frac{x_1 - x_0}{2} = \frac{h}{2}$ are the weights for trapezoidal integration. We already had presented this well-known quadrature, but the presented procedure also provides us with a systematic method to calculate its associated error.

$$E[f] = \int_{x_0}^{x_1} R_1(x)dx = \int_{x_0}^{x_1} \frac{(x^2 - (x_1 + x_0)x - x_1x_0)f''(\xi)}{2} dx$$

$$= \frac{f''(\xi)}{2}(x_1 - x_0)\left[\frac{x_1^2 + x_1x_0 + x_0^2}{3} - \frac{(x_1 + x_0)^2}{2} - x_1x_0\right]$$

$$= -\frac{f''(\xi)}{12}(x_1 - x_0)^3 \Rightarrow E[f] = -\frac{f''(\xi)}{12}(h)^3 = O(h^3)$$

The result is in agreement with the theorem presented for a quadrature with precision degree m, $E[f] = O(h^{m+2})$. This example shows that if one desires a quadrature with higher precision, one may use Lagrange polynomials to approximate the function by higher-order polynomials, leading to a quadrature with a greater precision, since $E[f] = O(h^{m+2})$.

There are many different quadratures with higher degree of precision to choose from, but the most commonly used are

Trapezoidal rule with precision degree $= 1$

$$\int_{x_0}^{x_1} f(x)dx = \frac{h}{2}[f(x_0) + f(x_1)] - \frac{f''(\xi)}{12}h^3$$

and Simpson rule with precision degree $= 3$

$$\int_{x_0}^{x_2} f(x)dx = \frac{h}{3}[f(x_0) + 4f(x_1) + f(x_2)] - \frac{f^{(4)}(\xi)}{90}h^5$$

5.4 Composite Quadrature

The integration rules presented in Section 5.3 were all for a small segment of a line. The reason being that if we wanted to approximated the function over interval $[a, b]$, as seen in Chapter 4, we need to use higher order polynomials to closely approximate the function near the nodal points. However, such an approximation leads to insta-bility and oscillation especially near the end points. To avoid this shortcoming, one breaks up the interval $[a, b]$ into M subintervals, and applies one of the quadratures developed in Section 5.3 to each of the subintervals. Such a procedure is known as composite quadrature rule. For example let $\gamma = \frac{b-a}{M}$ and $\tilde{x}_m = \tilde{x}_0 + m\gamma$, $m = 0, 1, \ldots, M$ and $\tilde{x}_0 = a$ & $\tilde{x}_M = b$. It then follows

$$I[f] = \int_a^b f(x)dx = \sum_{m=0}^{M-1} \int_{\tilde{x}_m}^{\tilde{x}_{m+1}} f(x)dx = \sum_{m=0}^{M-1} I_m[f]$$

where $I_m[f] = \int_{\tilde{x}_m}^{\tilde{x}_{m+1}} f(x)dx$.

Using a desired quadrature $Q_m[f]$ over the interval $[\tilde{x}_m, \tilde{x}_{m+1}]$ one replaces $I_m[f]$ by the chosen quadrature $Q_m[f]$ with J standing for one less than total number of nodal points needed to compute $Q_m[f] = \sum_{j=0}^J \tilde{w}_j f(x_j)$. In other words we subdivide the interval $[a, b]$ to $N = JM$ equally spaced nodes $x_j = a + jh$, with $h = \frac{b-a}{N} = \frac{\gamma}{J}$. Hence: $\tilde{x}_1 = a + \gamma = a + hJ = x_J$ and $\tilde{x}_m = a + m\gamma = a + mJh = x_{mJ}$, and

$$Q[f] = \sum_{m=0}^{M-1} Q_m[f] = \sum_{m=0}^{M-1} \sum_{j=0}^J \tilde{w}_j f(x_{(mJ+j)}) = \sum_{n=0}^N w_n f(x_n)$$

The weights for quadrature $Q_m[f]$ is denoted by $\tilde{w}_j$ and w_n is the corresponding weights for the composite quadrature $Q[f]$. From the above it follows:

$$I[f] = Q[f] + E[f], \text{ with } E[f] = \sum_{m=0}^{M-1} E_m[f]$$

As an example let us find the rule for composite Trapezoidal quadrature. By definition $J = 1$, $N = JM = M$, $\tilde{w}_j = \frac{h}{2}$, $\tilde{x}_j = x_j = a + jh$, & $h = (b-a)/N$.

$$Q_T[f] = \sum_{m=0}^{M-1} \sum_{j=0}^1 \tilde{w}_j f(x_{(mJ+j)}) = \frac{h}{2}[f(x_0) + f(x_1) + f(x_1) + f(x_2)$$

$$+ f(x_2) + f(x_3) + \cdots + f(x_{M-2}) + f(x_{M-1})$$

$$+ f(x_{M-1}) + f(x_M)]$$

$$= \frac{h}{2}(f(a) + f(b)) + h \sum_{m=1}^{M-1} f(x_m) = \sum_{n=0}^N w_n f(x_n)$$

Hence, $w_0 = w_N = h/2, w_m = h$ for $m = 1, 2, \ldots, (N-1)$, and truncation error

$$E_T[f] = -\sum_{m=0}^{M-1} \frac{f''(\xi_m)}{12}(h)^3 = -M\frac{f''(\xi)}{12}(h)^3 = -\frac{b-a}{h}\frac{f''(\xi)}{12}(h)^3$$

$$= -\frac{(b-a)f''(\xi)}{12}h^2, \text{ where } x_m \le \xi_m \le x_{m+1} \ \& \ a \le \xi \le b$$

Likewise for composite Simpson quadrature, by its definition $J = 2$, $N = JM = 2M, \tilde{w}_0 = \tilde{w}_2 = \frac{h}{3}, \tilde{w}_1 = \frac{4h}{3}$ and $h = \frac{b-a}{2M}$. Thus,

$$Q_S[f] = \sum_{m=0}^{M-1}\sum_{j=0}^{2} \tilde{w}_j f(x_{(mJ+j)}) = \frac{h}{3}[[f(x_0) + 4f(x_1) + f(x_2)]$$

$$+ [f(x_2) + 4f(x_3) + f(x_4)] + [f(x_4) + 4f(x_5) + f(x_6)]$$

$$+ \cdots + [f(x_{2M-4})] + [f(x_{2M-3}) + 4f(x_{2M-2})]$$

$$+ [f(x_{2M-2}) + 4f(x_{2M-1}) + f(x_{2M})]$$

$$= \frac{h}{3}(f(a) + f(b)) + \frac{4h}{3}\sum_{n=1}^{M} f(x_{2n-1}) + \frac{2h}{3}\sum_{n=1}^{M-1} f(x_{2n})$$

and

$$E_S[f] = -\sum_{m=0}^{M-1} \frac{f^{(4)}(\xi_m)}{90}h^5 = -M\frac{f^{(4)}(\xi)}{90}h^5 = -\frac{b-a}{2h}\frac{f^{(4)}(\xi)}{90}h^5$$

$$= -\frac{(b-a)f^{(4)}(\xi)}{180}h^4$$

Since the above results will be used often, for the convenience of the reader let us present a summary.

Composite Trapezoidal quadrature and its truncation error

$$Q_T[f] = \frac{h}{2}[f(a) + f(b)] + h\sum_{n=1}^{N-1} f(x_n), \quad E_T[f] = -\frac{(b-a)f''(\xi)}{12}(h)^2$$

where $x_n = a + nh$, and $h = \frac{b-a}{N}$.

Composite Simpson quadrature and its truncation error

$$Q_S[f] = \frac{h}{3}[f(a) + f(b)] + \frac{4h}{3} \sum_{n=1}^{M} f(x_{2n-1}) + \frac{2h}{3} \sum_{n=1}^{M-1} f(x_{2n})$$

$$E_S[f] = -\frac{(b-a)f^{(4)}(\xi)}{180}h^4, \quad x_n = a + nh, \quad h = \frac{b-a}{N} \quad \text{and } N = 2M$$

To give an example of above procedures let us integrate $f(x) = \sin x, x \in [0, \pi]$ numerically, using both composite Trapezoidal and composite Simpson rules.

$$\int_0^\pi \sin x\, dx = -\cos x\big|_0^\pi = 2$$

Since the exact value of the integral is known, this example will also allow us to test the found error estimate, using $2M = N = 20$.

MATLAB code for Table 5.1.

```
% inputting initial information
a = sin(0); b = sin(π); M = 10; N = 2 * M;
h = π/N; Nm = N - 1; S = h * (a + b)/2;
% finding Trapezoidal integral and its error
for j=1:Nm
x(j)=j*h; y(j)= sin(x(j)); S= S+h*y(j);
end;
E=2-S; Ef=-π * h²/12; SS=h*(a+b)/3;
% finding Simpson integral and its error
for n=1:M
SS=SS+ (4*h/3)*y(2*n-1);
end;
for n=1:(M-1)
SS=SS+ (2*h/3)*y(2*n);
end;
Esf=-π * h⁴/180; Es=2-SS;
```

As it can be seen from Table 5.1, the actual error in finding $I[f]$ is smaller than estimated error $E[f]$. This is due to the fact that for estimating $E[f]$ values of $f^{(2)}(\xi)$ & $f^{(4)}(\xi)$ were needed, but exact

Table 5.1. Finds $\int_0^\pi \sin(x)dx$ via Trapezoidal and Simpson rules.

Quadrature	N	h	Q[f]	Actual error	E[f]
Trapezoidal	20	0.1571	1.9959	0.0041	−0.0065
Simpson	20	0.1571	2.0000068	−0.0000068	0.0000106

value of ξ in general is not known. Thus max values of $f^{(2)}$ & $f^{(4)}$ were used to arrive at a conservative estimate of the errors.

There are many well-known quadratures with more accuracy, but for an introductory course in numerical analysis, the ability to apply Trapezoidal and Simpson quadratures are sufficient.

In finding definite integrals numerically, one needs to choose a desired value of subdivision N of interval $[a, b]$. A way to estimate N and consequently h is to decide what amount of error $E[f]$ is acceptable. That is to choose an ϵ and require $\epsilon \approx E[f]$. Then using known form of $E[f]$ for the desired quadrature one estimates N. As an example, take the composite Trapezoidal rule and note

$$E[f] = -\frac{(b-a)f''(\xi)}{12}(h)^2 = -\frac{(b-a)^3 f''(\xi)}{12N^2}$$

where $\xi \in [a, b]$. The above equation enables one to estimate N, the desired number of subdivisions.

$$N = \left\lceil \left| \frac{(b-a)^3 f''(\xi)}{12\epsilon} \right|^{\frac{1}{2}} \right\rceil$$

5.5 Adoptive Quadrature

If a function needing to be integrated behaves very differently over the interval of integration, then according to above presented estimate of N, value of nodal spacing h will be mainly determined by the region where f is changing more drastically. However, such an equal spacing methodology may not be the most efficient procedure. Adoptive Quadrature methodology allows nodal spacing to be different for different segments of the interval, where $f(x)$ is being integrated. To see the general idea used in this methodology consider a function $f(x)$ rapidly changing over $[a, x_1]$, but slowly changing over interval $[x_1, x_2]$ and moderately changing over the interval $[x_2, b]$. In such cases to have an efficient and accurate quadrature, one breaks the integration into segments. In the above example into three segments.

$$\int_a^b f(x)dx = \int_a^{x_1} f(x)dx + \int_{x_1}^{x_2} f(x)dx + \int_{x_2}^b f(x)dx$$

$$= I_1 + I_2 + I_3$$

Since in integral I_1, the integrand f is changing rapidly, one uses very small h. For I_2, f is changing slowly one uses larger h and for I_3 where f is changing moderately, one uses moderately small h. The reason being that the composite error $E[f] \approx f^{(m+1)}(\xi)h^{m+1}$ not only depends on h, but also on the value of $(m+1)$ derivative of f, where m depends on the quadrature rule being used. Hence, rapidly changing f having larger $f^{(m+1)}$ value and the need for smaller h to have same amount of error as compared with lesser changing f. These variations can be accommodated using the following procedure:

(i) Break the interval $[a, b]$ into J subsets, according to above mentioned behavior of $f(x)$. For example, $[a, b] = \cup_{j=1}^J [a_j, b_j)$ where $a_1 = a$, $b_J = b$, and define

$$I = \int_a^b f(x)dx = \sum_{j=1}^J \int_{a_j}^{b_j} f(x)dx = \sum_{j=1}^J I_j$$

The choice of nodal spacing h_j for each subinterval is then determined by following procedure, without finding derivatives of $f(x)$.

(ii) The desired accuracy ϵ is chosen, such that total error $E[f] \le \epsilon$, with error over each segment $[a_j, b_j]$ is selected to be

$$|E_j[f]| \le \epsilon_j = \frac{\epsilon(b_j - a_j)}{b - a}$$

(iii) To find $I_j[f] = \int_{a_j}^{b_j} f(x)dx$, for example using Simpson rule with ϵ_j error tolerance, one selects an N and finds $h = \frac{b_j - a_j}{N}$ for nodal spacing to find

$$I_j[f] = \int_{a_j}^{b_j} f(x)dx = Q_{S_h}[f] - \frac{(b_j - a_j)f^{(4)}(\xi_j)}{180}h^4$$

where $\xi_j \in (a_j, b_j)$. The above equation does not provide us with accurate information about the error, since $f^{(4)}(\xi_j)$ is not exactly known. To remedy this, one computes the quadrature again, using nodal spacing of $\frac{h}{2}$.

$$I_j[f] = \int_{a_j}^{b_j} f(x)dx = Qs_{\frac{h}{2}}[f] - \frac{(b_j - a_j)f^{(4)}(\tilde{\xi}_j)}{180}\left(\frac{h}{2}\right)^4$$

where $\tilde{\xi}_j \in (a_j, b_j)$. Equating above two equations

$$I_j[f] = Qs_h[f] - \frac{(b_j - a_j)f^{(4)}(\xi_j)}{180}h^4$$

$$= Qs_{\frac{h}{2}}[f] - \frac{(b_j - a_j)f^{(4)}(\tilde{\xi}_j)}{180}\left(\frac{h}{2}\right)^4$$

Assuming $f^{(4)}(\tilde{\xi}_j) \approx f^{(4)}(\xi_j)$ leads to

$$Qs_h[f] - Qs_{\frac{h}{2}}[f] \approx \left[1 - \frac{1}{16}\right]\frac{(b_j - a_j)f^{(4)}(\xi_j)}{180}h^4 = \frac{15}{16}E_j[f]$$

Hence,

$$|E_j[f]| \approx \frac{(b_j - a_j)|f^{(4)}(\xi_j)|}{180}h^4 \approx \frac{16}{15}|Qs_h[f] - Qs_{\frac{h}{2}}[f]|$$

Having estimated error over subinterval $[a_j, b_j]$ for a chosen h, if $E_j[f] \leq \epsilon_j$ then nodal spacing h is satisfactory, otherwise one needs to follow same procedure and reduce h by half again till desired error size is obtained. The outlined procedure is a practical method for finding nodal spacing h that makes estimated integral over a subinterval to have an error less than accepted tolerance, without having to know $f^{(4)}(\xi_j)$ when using composite Simpson rule. As an example, see Problem 5.6 for more details.

5.6 Singular/Improper Integrals

The rules developed to find a quadrature all require finding value of $f(x)$ at the nodal points x_n. The question then becomes how to modify developed methods, when the integrand is singular, but it is still integrable. For example, consider integrating

$$\int_{x_0}^{x_1} \frac{g(x)}{(x-a)^\alpha} dx, \text{ when } 0 < \alpha < 1, \ g(a) \neq 0, \ x_0 < a < x_1$$

and $g'(x)$ is finite in the neighborhood of a. Due to its singularity, the standard procedure to deal with such an integral is to break the integral at $x = a$

$$\int_{x_0}^{x_1} \frac{g(x)}{(x-a)^\alpha} dx = \int_{x_0}^{a} \frac{g(x)}{(x-a)^\alpha} dx + \int_{a}^{x_1} \frac{g(x)}{(x-a)^\alpha} dx = I_1 + I_2$$

The problems is now reduced to the cases when only at one end of the interval, the integrand is singular. Next one subtracts the singular part of the integrand and integrates it analytically. The remaining part which is not singular is evaluated applying a desired quadrature procedure.

$$I_1 = \int_{x_0}^{a} \frac{g(x) - g(a)}{(x-a)^\alpha} dx + \int_{x_0}^{a} \frac{g(a)}{(x-a)^\alpha} dx = I_3 + \frac{g(a)(x-a)^{1-\alpha}}{1-\alpha} \Big|_{x_0}^{a}$$

$$I_2 = \int_{a}^{x_1} \frac{g(x) - g(a)}{(x-a)^\alpha} dx + \int_{a}^{x_1} \frac{g(a)}{(x-a)^\alpha} dx = I_4 + \frac{g(a)(x-a)^{1-\alpha}}{1-\alpha} \Big|_{a}^{x_1}$$

Since the integrands for I_3 and I_4 are finite over the intervals of interest, they can be evaluated by reducing them to quadratures. Likewise, their errors $E[f]$ can also be estimated. The idea of subtracting an integrable function which is singular works well in many situations even if the function of interest is not singular over the region of interest. A concrete example is the integral

$$I = \int_{0.05}^{\frac{\pi}{2}} \frac{dx}{1 - \cos x} = -\cot \frac{x}{2} \Big|_{0.05}^{\frac{\pi}{2}} = 38.9917$$

The integrand over $[0.05, \frac{\pi}{2}]$ is not singular. Thus, we can evaluate it numerically using a quadrature. If Simpson Composite quadrature is used with error tolerance of 0.00001 we need to subdivide the interval to $N = 482$ parts, due to the drastic change in value of

$\frac{1}{1-\cos x}$, for values of x near 0.05. However, if we use Taylor formula and approximate $1 - \cos x \approx \frac{x^2}{2}$ near $x = 0$, and use $\frac{2}{x^2}$ as the function to be subtracted from the integrand

$$I = \int_{0.05}^{\frac{\pi}{2}} \left[\frac{1}{1 - \cos x} - \frac{2}{x^2} \right] dx + \int_{0.05}^{\frac{\pi}{2}} \frac{2}{x^2} dx = I_1 + I_2, \text{ where}$$

$$I_1 = \int_{0.05}^{\frac{\pi}{2}} \left[\frac{x^2 - 2 + 2\cos x}{x^2 - x^2 \cos x} \right] dx, \text{ and } I_2 = \int_{0.05}^{\frac{\pi}{2}} \frac{2}{x^2} dx = -\frac{4}{\pi} + \frac{2}{0.05}$$

We find for I_1, the value of the integrand does not change drastically for x near 0.05, and application of Simpson quadrature results in finding the integral with same tolerance, requires only two subdivisions of the interval. This example shows that one can find more efficient procedures when integrating a rapidly changing integrand, by modifying the integrand so it does not drastically change near singular points. Another related topic that makes use of similar procedure is evaluation of principal value integrals.

5.7 Principal Value Integrals

In many branches of applied sciences, one needs to evaluate integrals defined as principal value integrals. The symbol usually used to denote a principal value integral is P, which indicates the way limit of an integral is taken when approaching a singularity.

$$P \int_c^b f(x) dx \equiv \lim_{\epsilon \to 0} \left[\int_c^{a-\epsilon} f(x) dx + \int_{a+\epsilon}^b f(x) dx \right]$$

If both $\lim_{\epsilon \to 0} \int_c^{a-\epsilon} f(x) dx$ & $\lim_{\epsilon \to 0} \int_{a+\epsilon}^b f(x) dx$ exist, above definition for Principal Value integral gives the same result as the standard integral of f over $[c, b]$. However, if the limits don't exist individually, then above definition is a way to define an integral of $f(x)$ even when f(x) diverges as x tends to a. A simple example is $\int_{-1}^{1} \frac{1+x}{x} dx$. As we know this integral does not exist, since

$\int \frac{1}{x} dx = \ln |x|$ is divergent as $x \to 0$. However,

$$P \int_{-1}^{1} \frac{1+x}{x} dx = \lim_{\epsilon \to 0} \left[\int_{-1}^{-\epsilon} \frac{1+x}{x} dx + \int_{\epsilon}^{1} \frac{1+x}{x} dx \right]$$

$$= \lim_{\epsilon \to 0} \left[(x + \ln |x|)_{-1}^{-\epsilon} + (x + \ln |x|)_{\epsilon}^{1} \right]$$

$$= \lim_{\epsilon \to 0} [-\epsilon + 1 + \ln |\epsilon| - \ln |1|$$

$$+ 1 - \epsilon + \ln |1| - \ln |\epsilon|] = 2$$

In other words, although $\int_{-1}^{1} \frac{1+x}{x} dx$ does not exist, its principal value integral $P \int_{-1}^{1} \frac{1+x}{x} dx$ exists and is finite. This example makes it apparent that similar procedures developed for integration of singular integrands could be applied to evaluate principal value integrals. For example, let $f(x) = \frac{g(x)}{x-a}$, with $g(a) \neq 0$, then its principal value integral takes the form

$$P \int_{c}^{b} \frac{g(x)}{x - a} dx = \lim_{\epsilon \to 0} \left[\int_{c}^{a-\epsilon} \frac{g(x)}{x - a} dx + \int_{a+\epsilon}^{b} \frac{g(x)}{x - a} dx \right] \equiv I_1 + I_2,$$

where

$$I_1 = \int_{c}^{a-\epsilon} \frac{g(x)}{x - a} dx = I_3 + I_4, \quad I_2 = \int_{a+\epsilon}^{b} \frac{g(x)}{x - a} dx = I_5 + I_6,$$

$$I_3 = \int_{c}^{a-\epsilon} \frac{g(x) - g(a)}{x - a} dx, \quad I_4 = \int_{c}^{a-\epsilon} \frac{g(a)}{x - a} dx,$$

$$I_5 = \int_{a+\epsilon}^{b} \frac{g(x) - g(a)}{x - a} dx, \quad \text{and} \quad I_6 = \int_{a+\epsilon}^{b} \frac{g(a)}{x - a} dx.$$

Note

$$I_4 + I_6 = g(a)[\ln |-\epsilon| - \ln |c - a| + \ln |b - a| - \ln |\epsilon|].$$

Thus

$$\lim_{\epsilon \to 0} [I_4 + I_6] = g(a) [\ln |b - a| - \ln |c - a|], \quad \text{and}$$

$$P \int_{c}^{b} \frac{g(x)}{x - a} dx = \lim_{\epsilon \to 0} [I_3 + I_5] + g(a) \ln \left(\frac{b - a}{a - c} \right).$$

Let us note the integrand $\frac{g(x)-g(a)}{x-a}$ appears in both I_3 & I_5, and $\frac{g(x)-g(a)}{x-a} = g'(\xi)$, with ξ in the neighborhood of a. Hence, if g has a finite derivative in the neighborhood of "a," the integrands appearing in I_3 & I_5 will be finite and application of any standard numerical quadrature procedure, will enable us to find the desired integrals.

Another type of integrals that appears in applied sciences is the integration over an infinite interval $[a, \infty)$, where standard reduction to a quadrature does not work, since subdivision length $h = (\infty - a)/N = \infty$. A way to find such integrals numerically is to modify the problem.

5.8 Integration Over Infinite Intervals

Let us consider numerical integration of $f(x)$ over interval $(-\infty, \infty)$ assuming $f(x)$ is continuous over $(-\infty, \infty)$ and $f(x) = o(\frac{1}{|x|})$ as $|x|$ tends to infinity. The standard procedure to work with such an integral is to write it as

$$\int_{-\infty}^{\infty} f(x)dx = \int_{-\infty}^{-1} f(x)dx + \int_{-1}^{1} f(x)dx + \int_{1}^{\infty} f(x)dx = I_1 + I_2 + I_3$$

where

$$I_1 = \int_{-\infty}^{-1} f(x)dx, \quad I_2 = \int_{-1}^{1} f(x)dx, \quad \text{and } I_3 = \int_{1}^{\infty} f(x)dx$$

As it can be seen from the above, I_1 and I_3 are similar, i.e. if we change variable x to $-x$ we find $I_1 = \int_{-\infty}^{-1} f(x)dx = \int_{1}^{\infty} f(-x)dx$. The integral I_2 is over a finite interval and can be found numerically by reducing it to a quadrature. Thus, to study integration over infinite intervals, it is sufficient to present a numerical procedure for integrating a function over the interval $[1, \infty)$.

To see how to integrate $I_3 = \int_{1}^{\infty} f(x)dx$ we first change variable x to ρ satisfying the relation $x = \frac{1}{\rho}$. Thus, $dx = -\frac{d\rho}{\rho^2}$ and note

$$I_3 = \int_{1}^{\infty} f(x)dx = \int_{0}^{1} \frac{f(\frac{1}{\rho})}{\rho^2} d\rho$$

The above procedure reduces the integration over an infinite interval to integration over finite intervals. To show how I_3 integration can be carried out numerically, let us define $F(\rho) = \frac{f(\frac{1}{\rho})}{\rho^2}$ and recall we assumed $f(x) = o(\frac{1}{x})$, which implies as x tends to infinity, $f(x) = o(\frac{1}{|x|}) \sim \frac{c_0}{x^{1+\epsilon}} = c_0 \rho^{1+\epsilon}$, where $\epsilon > 0$ and c_0 is a finite constant. This assumption leads to $F(\rho) \sim \frac{c_0 \rho^{1+\epsilon}}{\rho^2} = \frac{c_0}{\rho^{1-\epsilon}}$, indicating $F(\rho)$ is not too singular as ρ approaches zero and the procedure we discussed on how to deal with singular integrals can be applied to numerically find I_3. The same reasoning also applies to I_1, thus $I = \int_{-\infty}^{\infty} f(x)dx$ can be found numerically, for $f(x)$ defined over $(-\infty, \infty)$ when $f(x) = o(\frac{1}{|x|})$ as $|x|$ tends to infinity.

Example 1. Let's recall a well-known definite integral often used for demonstration in different applied disciplines, that is

$$I = \int_0^\infty e^{-x^2} \cos 2\alpha x \, dx = \frac{\sqrt{\pi}}{2} e^{-\alpha^2}$$

Here, we demonstrate how to find I numerically using the above procedure.

$$I = \int_0^\infty e^{-x^2} \cos(2\alpha x)dx = I_1 + I_2,$$

where $I_1 = \int_0^1 e^{-x^2} \cos(2\alpha x)dx$ & $I_2 = \int_1^\infty e^{-x^2} \cos(2\alpha x)dx$

Next, I_2 is modified by changing variable x to ρ via the relation $x = \frac{1}{\rho}$ and find

$$I_2 = \int_0^1 \frac{e^{-\frac{1}{\rho^2}} \cos \frac{2\alpha}{\rho}}{\rho^2} d\rho$$

Since $e^{-\frac{1}{\rho^2}}$ goes to zero much faster than ρ^2 as ρ goes to zero, the function $\frac{e^{-\frac{1}{\rho^2}} \cos \frac{2\alpha}{\rho}}{\rho^2}$ goes to zero as ρ tends to zero. Thus, both I_1 and I_2 integrands are finite on $[0,1]$ and can be numerically integrated using standard methods. Numerical results for Example 1, is shown in Table 5.2.

Table 5.2. Numerical results for $\int_0^\infty e^{-x^2} \cos 2\alpha x \, dx$, when $\alpha = \frac{1}{2}$.

Quadrature	N	h	Q[f]	$I = \int_0^\infty e^{-x^2} \cos 2\alpha x \, dx$	Error= $I - Q[f]$
Simpson	20	0.05	0.6902	$\frac{\sqrt{\pi}}{2} e^{-\alpha^2}$, $\alpha = \frac{1}{2}$	$-5.8455e * 10^{-07}$

MATLAB code for Table 5.2

```
% inputting initial information
a=0.5;M=10; N=2*M; h=(1-0)/N; Ex=sqrt(pi)/2* exp(-a^ 2);
S1=(exp(0)+exp(-1)*cos(2*a))*h/3; S2=(0+exp(-1)*cos(2*a))*h/3;
% transforming variable and integrand
for n=1:N
s(n)=h*n; r(n)=1/s(n); g(n)=exp(-(s(n))^ 2)*cos(2*a*s(n));
f(n)= (exp(-(r(n))^ 2)*cos(2*a*r(n)))/((s(n))^ 2);
end;
for n=1:M
S1=S1+ (4*h/3)*g(2*n-1); S2=S2+ (4*h/3)*f(2*n-1);
end;
% finding the integral and its error
for n=1:(M-1)
S1=S1+ (2*h/3)*g(2*n); S2=S2+ (2*h/3)*f(2*n);
end;
ST=S1+S2; ET=Ex-ST;
```

Example 2. For the second example we select another well-known integral

$$I = \int_{-\infty}^{\infty} \frac{e^{\alpha x}}{e^x + 1} dx = \frac{\pi}{\sin \alpha \pi}, \quad 0 < \alpha < 1$$

For simplicity of the presentation let use choose $\alpha = \frac{1}{2}$ and find value of I numerically. Following the mentioned procedure

$$I = \int_{-\infty}^{-1} \frac{e^{\alpha x}}{e^x + 1} dx + \int_{-1}^{1} \frac{e^{\alpha x}}{e^x + 1} dx + \int_{1}^{\infty} \frac{e^{\alpha x}}{e^x + 1} dx = I_1 + I_2 + I_3$$

Changing variable x to $-x$ in I_1, we have

$$I_1 = \int_{-\infty}^{-1} \frac{e^{\alpha x}}{e^x + 1} dx = \int_{1}^{\infty} \frac{e^{-\alpha x}}{e^{-x} + 1} dx, \quad \text{and changing } x \text{ to } \rho = \frac{1}{x}$$

$$I_1 = \int_{0}^{1} \frac{e^{\frac{-\alpha}{\rho}}}{e^{\frac{-1}{\rho}} + 1} \frac{d\rho}{\rho^2}, \quad \text{and multiplying integrand of } I_3 \text{ by } \frac{e^{-x}}{e^{-x}}$$

$$I_3 = \int_1^\infty \frac{e^{\alpha x}}{e^x + 1} dx = \int_1^\infty \frac{e^{(-1+\alpha)x}}{1 + e^{-x}} dx = \int_1^\infty \frac{e^{-\frac{1}{2}x}}{e^{-x} + 1} dx = I_1$$

Taking $\xi = e^{\frac{-\alpha}{\rho}}$, $d\xi = \alpha \dfrac{e^{\frac{-\alpha}{\rho}} d\rho}{\rho^2}$

$$I_1 = I_3 = \int_0^1 \left[\frac{e^{\frac{-\alpha}{\rho}}}{\rho^2} \right] \frac{d\rho}{1 + e^{\frac{-1}{\rho}}} = \frac{1}{\alpha} \int_0^{e^{-\alpha}} \frac{d\xi}{1 + \xi^{\frac{1}{\alpha}}}$$

Since all the integrands are finite over the intervals of interest, let us use the Simpson quadrature to evaluate the integrals. Table 5.3 shows numerical results for Example 2.

Table 5.3.   Numerical result for integral $I = \int_{-\infty}^\infty \frac{e^{\alpha x}}{e^x+1} dx$ for $\alpha = \frac{1}{2}$.

Quadrature	N	h	$Q[f]$	$I = \int_{-\infty}^\infty \frac{e^{x/2}}{e^x+1} dx$	Error$=I - Q[f]$
Simpson	20	0.05	3.1416	π	$-1.5546 * 10^{-07}$

MATLAB code for Table 5.3.

```
% inputting initial information
a=1/2;M=10; N=2*M; h=(exp(-a)-0)/N; h2=(1+1)/N; Ex=(pi);
S2=(exp(-a)/(exp(-1)+1)+exp(a)/(exp(1)+1))*h2/3;
S3=(1+1/(1+(exp(-a))^ (1/a)))*h/3;
% defining relevant variables and integrands
for n=1:N
s2(n)=-1+h2*n; s3(n)=0+h*n;
f2(n)= exp(a*s2(n))/(exp(s2(n))+1);
f3(n)= 1/(1+(s3(n))^ (1/a));
end;
% finding relevant quadrature and associated error
for n=1:M
S2=S2+ (4*h2/3)*f2(2*n-1); S3=S3+ (4*h/3)*f3(2*n-1);
end;
for n=1:(M-1)
S2=S2+ (2*h2/3)*f2(2*n); S3=S3+ (2*h/3)*f3(2*n);
end;
ST=S2+(2/a)*S3; ET=Ex-ST;
```

In summary, above examples demonstrate that if the given definite integral exists, then by judicious change of variables and/or subtracting the singular term(s) from the integrand, one can modify the integral to be over a finite domain with finite integrand, so that quadrature procedures could be applied to find the integral numerically.

Exercises

5.1. Find a numerical difference approximation of $\frac{df(x)}{dx}|_{x=x_0}$, with truncation error of $O(h^4)$, where $h = x_{i+1} - x_i$.

5.2. Find a numerical difference approximation of $\frac{d^2 f(x)}{dx^2}|_{x=x_0}$, having truncation error $O(h^4)$, where $h = x_{i+1} - x_i$.

5.3. (a) Apply Richardson extrapolation method to the following approximation:

$$\frac{df(x)}{dx} = \frac{f(x+h) - f(x-h)}{2h} + O(h^2)$$

(b) Apply your results to

$$f(x) = e^x + 3x^2$$

and see if applying Richardson extrapolation method makes any improvement in finding $f'(0)$ for $h = 0.2$ and $\alpha = 2$ as defined in Section 5.2.

5.4. (a) Apply Richardson extrapolation method to the following approximation:

$$\frac{d^2 f(x)}{dx^2} = \frac{f(x+h) - 2f(x) + f(x-h)}{h^2} + O(h^2),$$

where $h = x_{i+1} - x_i$

(b) Apply your results to $f(x) = \cos x^2$ and see if applying Richardson extrapolation method makes any improvement in finding $f''(0)$ for $h = 0.2$ and $\alpha = 2$ as defined in Section 5.2 for the above.

5.5. Let $I[f] = \int_{x_0}^{x_2} f(x)dx = Q[f] + E[f]$. Choose $Q[f]$ to be the Simpson quadrature

$$Q[f] = \frac{h}{3}[f(x_0) + 4f(x_1) + f(x_2)]$$

(a) Find the precision of Simpson's quadrature $Q[f]$.

(b) Prove that the error $E[f] = -\frac{h^5}{90} f^{(4)}(\xi)$.

5.6. Integrate $\int_0^2 e^{-x^2} dx$, using composite trapezoidal rule. Use the smallest N that will make the absolute error $E_T \le 10^{-5}$.

5.7. (a) Let $f(x) \in C^2[a, b]$. Find numerical error in finding $\int_a^b f(x)dx$ using composite trapezoidal rule by evaluating its quadrature first for $h = \frac{b-a}{M}$, and then finding the associated quadrature for $\frac{h}{2}$ to better estimate $f''(\xi)$, where M is an integer and $a \le \xi \le b$.

(b) Use the error estimate you found in Problem 5.7(a) to find error for numerically finding $I = \int_0^2 e^{-x^2} dx$, using the composite trapezoidal rule with $M = 10$.

5.8. (a) Find $I = \int_0^1 e^{-10x} dx = Q_{[0,1]}[f] + E_{[0,1]}[f]$ using the composite Simpson rule with $N = 16$ and $h = \frac{1}{N}$. By comparing found value of $Q_{[0,1]}[f]$ and the exact value of I, find the $E[f]$.

(b) Since change of $f(x) = e^{-10x}$ is more rapid over $0 \le x \le 0.25$, write $I = I_1 + I_2$ with $I_1 = \int_0^{0.25} f(x)dx = Q_{[0,0.25]}[f] + E_{[0,0.25]}[f]$ and $I_2 = \int_{0.25}^1 f(x)dx = Q_{[0.25,1]}[f] + E_{[0.25,1]}[f]$. Integrate I_1 using composite Simpson rule with $\tilde{h} = \frac{h}{2}$. To find I_2 use composite Simpson rule with the same h as used in part (a). Compare $Q_{[0,0.25]}[f] + Q_{[0.25,1]}[f]$ with the analytically found value of I to find $E_{[0,1]}[f]$.

(c) From quantities computed in parts (a) and (b), compare estimated error $|E_{[0, 0.25]}[f]| \approx \frac{16}{15}|Q_{S_h}[f] - Q_{S_{\frac{h}{2}}}[f]|$ with error made in finding I_1 numerically.

5.9. Same as Problem 5.7(b), except apply composite Simpson rule for integration and find the error estimate $E_S[f]$ using $Q_{S_h}[f]$ and $Q_{S_{\frac{h}{2}}}[f]$, for the same h as in Problem 5.7(b).

5.10. Use composite trapezoidal rule with $M = 10$ to numerically find $\int_0^1 \frac{e^{2x}}{\sqrt{1-x^2}} dx$, by subtracting appropriate function from the integrand to make it non-singular.

5.11. Use Simpson rule to find $\int_0^1 \frac{e^{2x}}{\sqrt{1-x^2}} dx$ numerically with $N = 16$, by first using the transformation $x = \sin\theta$.

5.12. Find $P \int_0^4 \frac{\cos \pi x}{x^2-4} dx$ using composite Simpson rule with $N = 18$.

5.13. You have seen how to find $I = \int_0^\infty e^{-x^2} dx$ analytically. Now, use numerical procedure developed for infinite intervals to find I numerically, by using Simpson's quadrature with $N = 20$. Compare estimated error and the actual error for the found quadrature.

5.14. Evaluate $\int_0^\infty \frac{2x^4 - x^2}{e^x - 1} dx$ by using composite trapezoidal rule with $N = 10$, and numerically estimate the error.

5.15. Evaluate $\int_0^\infty \frac{e^{-x}}{\sqrt{x}} dx$ using composite trapezoidal rule with $N = 20$ and numerically estimate the error.

Chapter 6

Initial Value Problems

Numerical methods for finding solutions to differential equations is of immense practical importance in theoretical and applied sciences. To motivate its study let's mention Newton's gravitational laws that form the foundation of classical mechanics and Maxwell's equations that govern electromagnetism. To be more specific consider an object with a mass m falling under the force of gravity $F = mg$ and the air exerting a friction force $m\mu\frac{dx}{dt}$ on it, where μ is called coefficient of friction, and x is the object's displacement. According to Newton's second law of motion, we have

$$m\frac{d^2x}{dt^2} + m\mu\frac{dx}{dt} = mg = F$$

To find $x(t)$ we also need to specify the initial state of the object when it started to feel these forces. In other words location of the object x_0 at starting time t_0 and its velocity $v_0 = \frac{dx}{dt}$ at t_0. Such a problem is called a second-order ordinary differential equation (ODE) initial value problem.

To develop a general method that is also applicable to higher-order ODE's, let us write the general form of an nth-order ODE as

$$x^{(n)} = w(t, x, x', \ldots, x^{(n-1)})$$

where $x^{(n)} \equiv \frac{d^n x(t)}{dt^n}$, and function w depends on t and on $x^{(m)}$, with $m = 0, 1, 2, \ldots, (n-1)$.

As an example let us write the above mentioned free falling object using this notation

$$x'' = w(t, x, x') = g - \mu x'$$

It turns out to be more convenient if we write the nth-order ODE as a first-order ODE in a vector form. To achieve this one start with the following definition:

$$X(t) \equiv \begin{bmatrix} x(t) \\ x'(t) \\ \cdots \\ x^{(k)}(t) \\ \cdots \\ x^{(n-1)}(t) \end{bmatrix} \equiv \begin{bmatrix} x_1(t) \\ x_2(t) \\ \cdots \\ x_{k+1}(t) \\ \cdots \\ x_n(t) \end{bmatrix}$$

Thus,

$$X'(t) = \begin{bmatrix} x'(t) \\ x''(t) \\ \cdots \\ x^{(k+1)}(t) \\ \cdots \\ x^{(n)}(t) \end{bmatrix} = \begin{bmatrix} x_2(t) \\ x_3(t) \\ \cdots \\ x_{k+2}(t) \\ \cdots \\ w(t, X) \end{bmatrix} = \begin{bmatrix} F_1(t, X) \\ F_2(t, X) \\ \cdots \\ F_{k+1}(t, X) \\ \cdots \\ F_n(t, X) \end{bmatrix} = F(t, X)$$

where $w(t, X) \equiv w(t, x, x', \ldots, x^{(n-1)})$, $F_k(t, X) = x_{k+1}$ for $k = 1, 2, \ldots, (n-1)$, and $F_n(t, X) = w(t, X)$.

Using this vector representation of nth-order ODE, numerical methods developed for scalar first order ODE $x' = f(t, x)$ with $x(t_0) = x_0$ can be in general extended, using the above system of equations associated with the nth-order ODE

$$X' = F(t, X), \text{ where } X(t_0) = [x_0, x_0', \ldots, x_0^{(n-1)}]^\top$$

This observation motivates presenting numerical results mainly for first order scalar ODE initial value problem (IVP).

$$x' = f(t, x), \quad x(t_0) = x_0$$

Its extension to higher order ODE, usually follows using the above representation of higher order ODE as a system of first order ordinary differential equations.

6.1 Euler's Method

The first question to be looked at before finding numerical solution of an IVP is the existence of its solution. We will use Euler's method to show existence and find its numerical approximate solution. To show existence one needs $f(t, x)$ to be continuous over an open connected set $D \subset R^2$, with the initial data $(t_0, x_0) \in D$. Then a rectangle $S_0 \subset D$ as shown in Fig. 6.1 is defined

$$S_0 = \{(t, x) : |t - t_0| \le a \ \& \ |x - x_0| \le b\}$$

To further localize the region where possible solution could reside, let us note that the solution needs to satisfy $x' = f(t, x)$. That is

$$|x'| \le M, \ \text{with}, \ M = \max |f(t, x)|, \quad \forall (t, x) \in S_0$$

In other words, the region between the lines with slopes $\pm M$, shown in Fig. 6.1 passing through the initial data (t_0, x_0) will define the region where possible solution can reside. However, we also need to make sure the solution stays in S_0. To achieve this, a constant

$$c = \min \left(a, \frac{b}{M} \right)$$

is defined, which will provide us with two distinct possible cases of interest.

Case (a), where $c = a < \frac{b}{M}$,
Case (b), where $c = \frac{b}{M} < a$.

If case (a) is true, the lines with slopes $\pm M$ will intercept the vertical sides of S_0, (Fig. 6.1(a)), and S_0 is not modified, since the solution will remain in S_0. On the other hand, if case (b) is true, the lines with slopes $\pm M$ will intercept the horizontal sides of the rectangle S_0, (Fig. 6.1(b)), and the solution may leave the region S_0. In this case, S_0 is modified by defining a new rectangle

$$S = \{(t, x) : |t - t_0| \le c, \ |x - x_0| \le b\}$$

In order to keep the labeling uniform for what is to follow, for both cases, the region of interest will be labeled as S, even when S_0 is not modified, with the understanding that S is the same as S_0 for for

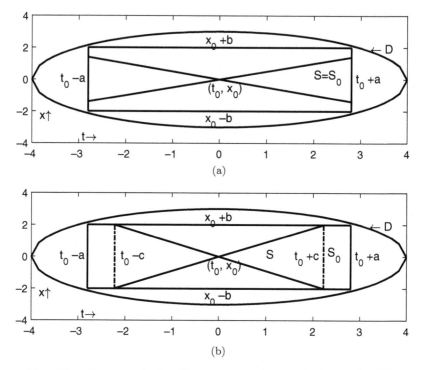

Fig. 6.1. Domain selection for an ϵ-approximate solution to the IVP.

case (a). Figure 6.1 shows the above procedure, where for $c = a$, the new rectangle S remains the same as $S_0 \subset D$, and for case $c < a$, the set S_0 is changed to set $S \subset S_0$.

Theorem. *For $x' = f(t, x)$ with $x(t_0) = x_0$, if $f(t, x)$ is continuous on the domain D, then for any $\epsilon > 0$ exist an ϵ-approximate solution satisfying*

$$|x'(t) - f(t, x(t))| < \epsilon, \text{ for } t \in [t_0, t_0 + c], \text{ where } c = \min(a, b/M)$$

Proof. Since $f(t, x)$ is uniformly continuous on S, for any $\epsilon > 0$ exists a $\delta > 0$ such that $|f(t, x) - f(t_0, y)| < \epsilon$ if $|t - t_0| < \delta$ & $|x - y| < \delta$. To prove existence of an ϵ-approximate solution, choose an integer J, as the number of subdivisions of the interval $[t_0, t_0 + c] \ni t_j = t_0 + jh$, where $h = \frac{c}{J}$ & $j = 0, 1, 2, \ldots, J$. Take J large enough so that $|t_{j+1} - t_j| < \min[\delta, \frac{\delta}{M}]$. Then the Euler's approximate solution is defined as

$$x(t) = x_j + f(t_j, x_j)(t - t_j), \quad t \in [t_j, t_{j+1}]$$

To show this is an ϵ-approximate solution, for $t_j \le t \le t_{j+1}$ one notes

$$|x(t) - x(t_j)| = |f(t_j, x_j)||t - t_j| \le M * \min \left[\delta, \frac{\delta}{M} \right] = \delta$$

and since Euler's formula approximates $x'(t) = x_j' + f(t_j, x_j)(t - t_j)' = 0 + f(t_j, x_j)$, for $t \in [t_j, t_{j+1}]$, it follows:

$$|x'(t) - f(t, x)| = |f(t_j, x_j) - f(t, x)| < \epsilon$$

since both $|t - t_j| \le \delta$ and $|x(t) - x_j| \le \delta$. It should be noted that in the above presentation $x(t_j)$ denotes the exact and x_j denotes the Euler's approximate value of the solution at t_j.

From the definition of ϵ-approximate solution, it follows that Euler's solution is an ϵ-approximate solution. Using Euler ϵ-approximate solution and letting ϵ tend to zero leads to the following existence theorem.

Cauchy–Peano Existence Theorem. If $f(t, x)$ is continuous on the defined rectangle S, then exists a solution $\phi(t)$ of $x' = f(t, x)$ on $|t - t_0| \le c$ satisfying the initial condition $\phi(t_0) = x_0$.

It can also be proven that if $f(t, x)$ is Lipschitz, with respect to x, then the solution to this IVP is unique.

Note. $f(t, x)$ is called Lipschitz respect to x, if there exists a finite constant L such that $|f(t, x) - f(t, y)| \le L|x - y|$ for all points $(t, x) \,\&\, (t, y) \in S$. A sufficient condition for $f(t, x)$ to be Lipschitz is to have

$$|f_x| = \left| \frac{\partial f(t, x)}{\partial x} \right| < \infty, \quad \forall \, (t, x) \in S.$$

Above sufficiency condition for Lipschitz is proven by using mean value theorem

$$\frac{f(t, \beta) - f(t, \alpha)}{\beta - \alpha} = \left. \frac{\partial f(t, x)}{\partial x} \right|_{x = \gamma}, \quad \text{where } \gamma \in [\alpha, \beta]$$

Thus, if $f_x(t, \gamma)$ is finite for all $(t, \gamma) \in D$ then

$$|f(t, x) - f(t, y)| = |f_x(t, \gamma)(x - y)| \leq L|x - y|,$$
$$\text{where } L = \max f_x(t, \gamma) \; \forall \; (t, \gamma) \in D$$

We should also note that Euler's approximate solution is a set of line segments joining neighboring nodal points. In other words, to find Euler's approximate solution, we only need to find x_j which is the approximate value of $x(t_j)$ for $j = 1, 2, \ldots, J$. This is achieved recursively, starting with the known initial value of $x(t_0) = x_0$, and using Euler's equation

$$x_{j+1} = x_j + f(t_j, x_j)h, \quad j = 0, 1, 2, \ldots, J, \; t_j = t_0 + jh \; \& \; h = \frac{c}{J}$$

Except for some simple ODE's, in general one cannot find a closed form solution to IVP. Thus, numerical methods are used extensively to find solutions to the initial value problems. The Euler's procedure forms the foundation for these numerical methods. It should also be noted that Lipschitz condition plays an essential role in numerical calculations, since the procedure used may not be able to find all the solutions. To see this better let's consider the following IVP.

$$x' = f(t, x) = 1.5x^{\frac{1}{3}}, \quad \text{with } x(0) = 0$$

This example does not satisfy Lipschitz condition, since $f_x = 0.5x^{-\frac{2}{3}}$ does not remain finite as x goes to 0. Hence, $f(t, x)$ is not Lipschitz for IVP with $x(0) = 0$. It is well known that this IVP has infinitely many solutions, but for the sake of demonstration let us present only two of these solutions. One solution being $x(t) = 0 \; \forall \; t \in [0, c]$. The other being $x = t^{\frac{3}{2}} \; \forall \; t \in [0, c]$, since $x' = \frac{3}{2}t^{\frac{1}{2}}$, and $t^{\frac{1}{2}} = [t^{\frac{3}{2}}]^{\frac{1}{3}} = x^{\frac{1}{3}}$. Thus $x = t^{\frac{3}{2}}$ is another solution of $x' = 1.5x^{\frac{1}{3}}$ that also satisfies the initial condition $x(0) = 0$. However, if we apply Euler's method

$$x(t) = x_j + f(t_j, x_j)(t - t_j) \text{ for } t \in [t_j, t_{j+1}], \; \& \; x(0) = x_0 = 0$$

one finds $f(0, 0) = 1.5x^{\frac{1}{3}}|_{x=0} = 0$, $x(t) = 0$ for $t \in [0, t_1]$. Thus, $x_1 = x(t_1) = 0$. Similarly all other $x_j = x(t_j)$ will also be zero. In other words, Euler's method gives us only the trivial solution

$x(t) = 0$, but knowing that the problem has more than one solution, allows us to search for other solutions, by appealing to the continuous dependence of the solution on the initial data. For example by trying to change the initial value a little, i.e. $x(0) = 0.001$ and see if Euler's method provides us with a non-trivial solution, close to the expected solution $x = t^{\frac{3}{2}}$. As it can be seen from Fig. 6.2, Euler's method does recover the second solution well, if we change the initial date a little to have $f(t, x)$ become Lipschitz in the neighborhood of the initial data. Furthermore, the continuity property of the solution respect to the initial data makes it possible to approximately find the second solution via Euler's method.

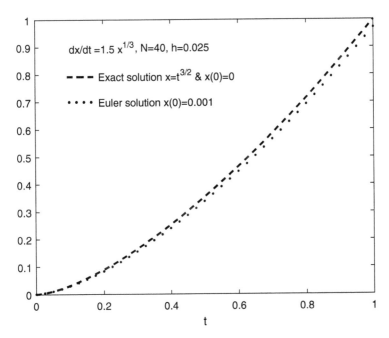

Fig. 6.2. Euler's solution when initial value is modified.

MATLAB code for Fig. 6.2

```
% inputting initial information
N=40; h= 1/N; x(1)=0.001; N1=N+1;
for m=1:N1
s(m)= h*(m-1); xe(m)= (x(m))^3/2; end;
```

% finding Euler's solution
for n=1:N
x(n+1)= x(n)+ 1.5*h* (x(n))^1/3; end
plot (s, x, 'k–', s,xe, 'k.', 'LineWidth', 1.4)

To estimate the error of Euler's method for a J number of subdivisions, let us again appeal to the Taylor's theorem

$$x(t) = x(t_j) + x'(t_j)(t - t_j)) + \frac{x''(\xi_j)}{2}(t - t_j)^2,$$

$$\text{for } t \in [t_j, t_{j+1}]$$

$$\text{or} \quad x(t_{j+1}) = x(t_j) + f(t_j, x(t_j))h + \frac{x''(\xi_j)}{2}h^2,$$

$$\text{where } \xi_j \in [t_j, t_{j+1}]$$

and $h = t_{j+1} - t_j = \frac{c}{J}$. This shows the truncation error in each step of Euler's method is

$$\text{Local error} = \frac{x''(\xi_j)}{2}h^2 \equiv O(h^2)$$

and error in carrying out J steps to find the solution over $[t_0, t_0 + c]$ is called

$$\text{Global error} = \sum_{j=1}^{J} \frac{x''(\xi_j)}{2}h^2 = \frac{Jh^2}{2}\sum_{j=1}^{J}\frac{x''(\xi_j)}{J} = \frac{cx''(\xi)}{2}h = O(h)$$

where $x''(\xi) = \sum_{j=1}^{J}\frac{x''(\xi_j)}{J}$ represents the average values of $x''(\xi_j)$ over the interval $[t_0, t_0+c]$. It is evident that the above results assumes $f_t(t, x)$ & $f_x(t, x)$ exist and finite over the domain of interest. Figure 6.2 also shows the difference between exact solution and Euler's solution at $t = 1$ is $0.0287 \sim h = 0.025$, which is consistent with the found global error estimate $O(h)$.

6.2 Extensions of Euler's Method

Section 6.1 showed that the Euler's method is based on first-term Taylor expansion of the solution over subintervals, and use of the line segment slope at $x(t_j)$ allows one to predict value of $x(t_{j+1})$. A natural extension of the Euler's method is to use 2-term Taylor expansion of the solution

$$x(t_{j+1}) - x(t_j) = x'(t_j)h + x''(t_j)\frac{h^2}{2} + \frac{x'''(\xi_j)}{3!}h^3$$

since $x''(t)$ can be found by taking total derivative of $\frac{dx(t)}{dt} = f(t, x(t))$ with respect to t, results in

$$\frac{d^2x(t)}{dt^2} = \frac{df(t, x(t))}{dt} = f_t(t, x) + f_x(t, x)x' = f_t(t, x) + f_x(t, x)f(t, x)$$

Such an extension of the Euler's method, results in a more accurate finite difference solution to $x' = f(t, x)$ with smaller truncation error. That is

$$x_{j+1} = x_j + f(t_j, x_j)h + [f_t(t_j, x_j) + f_x(t_j, x_j)f(t_j, x_j)]\frac{h^2}{2} + O(h^3)$$

Similarly by using more terms of Taylor series, the procedure can be continued to find more accurate finite difference representation of the solution to IVP $x' = f(t, x)$ by using Taylor expansion. However, such procedures require finding higher-order partial derivatives of $f(t, x)$ with respect to t and x, that becomes too tedious to find. To avoid direct computation of partial derivatives of $f(t, x)$, the idea of Runge-Kutta method is to estimate needed partial derivatives by making use of Taylor expansion formula for a function of two variables, $f(t, x)$, and evaluate f for values of the variables in the neighborhood of (t, x). For example, for the above 2-term Taylor series Runge-kuta method selects the related coefficients so to minimize the truncation error for second-term Runge-Kuta approximation. This leads to following RK2 procedure:

$$x_{j+1} = x_j + \frac{h}{4}\left[f(t_j, x_j) + 3f\left(t_j + \frac{2}{3}h, x_j + \frac{2}{3}hf(t_j, x_j)\right)\right] + O(h^3)$$

See A. Ralston for more details of the above RK2 method. To see how well RK2 method approximates the second-term Taylor expansion solution, the following numerical example is presented in Fig. 6.3.

$$\frac{dx}{dt} = 1 + x^2 \text{ with initial condition } x(0) = 0$$

Since $f(t, x)$ takes the form $f(t, x) = 1 + x^2$ for this example, Taylor 2-term procedure takes the form

$$x_{j+1} = x_j + f(t_j, x_j)h + x_j f(t_j, x_j)h^2 + O(h^3)$$

Figure 6.3, shows the difference between RK2 solution and Taylor second-term solution is consistent with the expectation that global error for the difference is $O(h^2)$, and here $h = 0.025$ & $h^2 = 0.000625$.

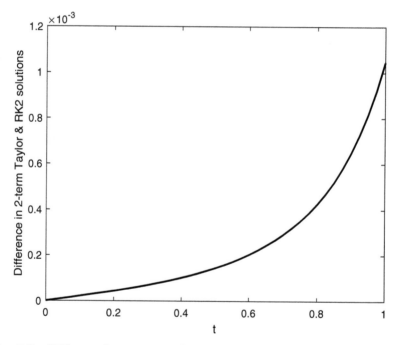

Fig. 6.3. Difference between second-term Taylor solution and RK2 solution for $\frac{dx}{dt} = 1 + x^2$, $x(0) = 0$, $N = 40$, and $h = \frac{1}{N}$.

MATLAB code for Fig. 6.3

```
% inputting initial information
N=40; h= 1/N; xr(1)=0;yt2(1)=0;
```

```
% finding 2-term Taylor and RK2 solutions
for m=1:N
x(m)= h*(m-1); ft2=1+(yt2(m))²;
fr2=xr(m)+(2/3)*h*(1+(xr(m) )²);
xr(m+1)=xr(m)+(h/4)*(1+(xr(m) )²+3*(1+(fr2)²));
yt2(m+1)=yt2(m)+h*ft2+yt2(m)*ft2*h²; end
% finding difference between 2-term Taylor and RK2 solutions
x(N+1)=1;yrk2=xr-yt2; plot ( x, yrk2,'k-','LineWidth', 1.3)
```

The most frequently used RK method is the classical Runge-Kutta method called RK4 which uses same idea as developed for RK2, but approximates the fourth-term Taylor expansion solution. It estimates values of x'', x''' and x'''' derivatives by finding $f(t,x)$ values in neighborhood of (t,x) which leads to following procedure

$$x_{k+1} = x_k + \frac{h}{6}[f_1 + 2f_2 + 2f_3 + f_4] + O(h^5)$$

where

$$f_1 = f(t_k,\ x_k), \quad f_2 = f\left(t_k + \frac{h}{2},\ x_k + \frac{h}{2}f_1\right)$$

$$f_3 = f\left(t_k + \frac{h}{2},\ x_k + \frac{h}{2}f_2\right), \quad f_4 = f(t_k + h,\ x_k + hf_3)$$

Numerical solution to IVP $x' = 1 + x^2$ with $x(0) = 0$, using RK4 and its exact solution $x(t) = \tan(t)$ are shown in Fig. 6.4.

MATLAB code for Fig. 6.4

```
% inputting initial information
N=40; h= 1/N; hh=h/2; y(1)=0; y1(1)=0;
% computing exact and RK4 solutions
for m=1:N
x(m)= h*(m-1); ye(m)= tan(x(m)); y(m+1)= y(m)+ h*(1+
(y(m))²);
f1= 1+(y1(m))²; f2=1+(y1(m)+hh*f1)²; f3=1+(y1(m)+hh*f2)²;
f4=1+(y1(m)+h*f3)²; y1(m+1)= y1(m)+ h*(f1+2*f2+2*f3+f4)/6;
end;
```

% finding difference between exact and RK4 solutions
x(N+1)=1; ye(N+1)=tan(x(N+1)); drk = ye-y1;
Figure(1); hold off;
plot (x,ye,'k–', x, y1,'ko','LineWidth', 1.3)
hold on; Figure(2); hold off;
plot (x, drk,'k.','LineWidth', 1.3)
hold on;

By construction RK4's local error for each step is $O(h^5)$ and global error is $O(h^4)$. Since for this example the agreement between exact and numerically computed solution via RK4 is of order 10^{-9}, in order to show the deviation between exact and its RK4 solution, it became necessary to present in Fig. 6.5, this difference between exact and its RK4 solution for the IVP of $\frac{dx}{dt} = 1 + x^2$ with $x(0) = 0$.

For the sake of more efficiency for large computations when using RK4, Fehlberg extended RK4 by estimating spacing h that makes error to be within a given tolerance at each step of calculation.

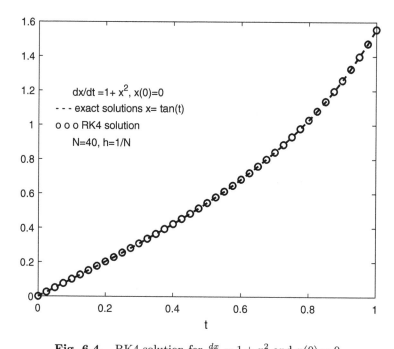

Fig. 6.4. RK4 solution for $\frac{dx}{dt} = 1 + x^2$ and $x(0) = 0$.

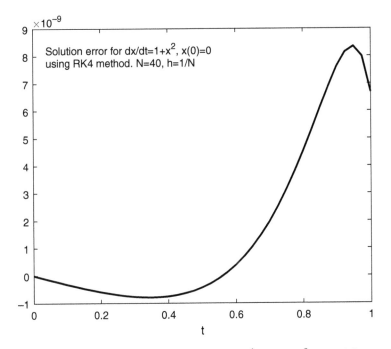

Fig. 6.5. RK4 error compared to exact for $\frac{dx}{dt} = 1 + x^2$ and $x(0) = 0$.

Fehlberg achieved this by applying both RK4 and RK5 to estimate the needed h. For this reason Fehlberg method is denoted as RKF45. The idea of RKF45 method is to estimate needed h as calculations being performed and not wait till solution $x(t)$ for all needed t are found before deciding if h was too large, which may necessitate the need to start calculation over again with a smaller h. The general idea of Fehlberg method for numerical solution of the initial value problem

$$x'(t) = f(t, x(t)), \quad x(t_0) = x_0, \ \& \ t_n = t_0 + nh$$

is to apply RK4 method to find an approximation of the exact solution $x(t_n)$ and denote it as x_n. In other words

$$x_n = x(t_n) + O(h^4) = x(t_n) + D_4 h^4$$

Similarly another approximate solution to above IVP is found by applying RK5 and denoting this approximate solution as $\tilde{x}_n \approx x(t_n)$.

That is apply the following procedure to find RK5 version of the solution to above IVP

$$\tilde{x}_{n+1} = \tilde{x}_n + \frac{16}{135}k_1 + \frac{6656}{12825}k_3 + \frac{28561}{56430}k_4 - \frac{9}{50}k_5 + \frac{2}{55}k_6$$

where

$$k_1 = hf(t_n, \tilde{x}_n)$$

$$k_2 = hf\left(t_n + \frac{1}{4}h, \tilde{x}_n + \frac{1}{4}k_1\right)$$

$$k_3 = hf\left(t_n + \frac{3}{8}h, \tilde{x}_n + \frac{3}{32}k_1 + \frac{9}{32}k_2\right)$$

$$k_4 = hf\left(t_n + \frac{12}{13}h, \tilde{x}_n + \frac{1932}{2197}k_1 - \frac{7200}{2197}k_2 + \frac{7296}{2197}k_3\right)$$

$$k_5 = hf\left(t_n + h, \tilde{x}_n + \frac{439}{216}k_1 - 8k_2 + \frac{3680}{513}k_3 - \frac{840}{4104}k_4\right)$$

$$k_6 = hf\left(t_n + \frac{1}{2}h, \tilde{x}_n - \frac{8}{32}k_1 + 2k_2 - \frac{3544}{2565}k_3 + \frac{1859}{4104}k_4 - \frac{11}{40}k_5\right)$$

The truncation error for RK5 method is of order h^5. That is

$$\tilde{x}_n = x(t_n) + O(h^5) = x(t_n) + D_5 h^5$$

Subtracting solution found using RK4 from above equation leads to

$$|\tilde{x}_n - x_n| = |D_5 h^5 - D_4 h^4| = D h^4$$

We have assumed h is sufficiently small, so that one can justify $D_5 h^5 - D_4 h^4 = O(h^4)$. In order to estimate desired step size h, one selects an error tolerance value, denoted by "TOL," so that $\epsilon \equiv \frac{|\tilde{x}_n - x_n|}{h} \leq \text{TOL}$.

Next, let h be changed to "$\hat{h}$" for the next step. Denote $\hat{h}$ by gh, where g is the factor that changes h to $\hat{h}$. That is

$$\epsilon = |\tilde{x}_n - x_n|/h = \left(\frac{D}{h}\right)h^4 \leq \text{TOL}\,\frac{h^4}{\hat{h}^4} = \frac{\text{TOL}}{g^4}$$

Finding g will allow us to define the next step size $\hat{h} = gh$ satisfying the tolerance requirement

$$g \leq \left(\frac{\text{TOL}}{\epsilon}\right)^{\frac{1}{4}} = \left(\frac{\text{TOL} * \text{h}}{|\tilde{x}_n - x_n|}\right)^{\frac{1}{4}}$$

In order to have a conservative and a systematic correction to the step size h, the following form is commonly used

$$g = \left(\frac{\text{TOL} * \text{h}}{2|\tilde{x}_n - x_n|}\right)^{\frac{1}{4}} \approx 0.84 \left(\frac{\text{TOL} * \text{h}}{|\tilde{x}_n - x_n|}\right)^{\frac{1}{4}}$$

where x_n and $\tilde{x}_n$ are the calculated values of the solution at t_n for the step size h, using RK4 and RK5 methods, respectively. The above procedure allows step size h to be updated to a new step size h, as computation using RKF45 method is in progress. Such a procedure will take more computer time since solution via RK4 and RK5 need to be calculated for finding the new step size, but RKF45 is very useful for large computations that requires selecting h in an efficient way that also allows computation error to be within an acceptable range, by changing the step size as needed throughout the procedure.

6.3 Heun's Method

Heun developed another approach for finding numerical solution to initial value problem $x' = f(t, x)$ with $x(t_0) = x_0$. Unlike Euler that chose numerical differentiation to solve $x' = f(t, x)$, Heun chose numerical integration to solve the same problem. He made use of the integral representation of the given initial value problem.

$$x(t) = x_0 + \int_{t_0}^{t} f(s, x(s))ds$$

By differentiation of above equation, one can show solution to above integral equation satisfies $x(t_0) = x_0$ and $x'(t) = f(t, x(t))$.

Next, using trapezoidal integration with nodal points at $t_j = t_0 + jh$, one finds

$$\int_{t_0}^{t_1} f(s, x(s))ds = \frac{h}{2}[f(t_0, x(t_0)) + f(t_1, x(t_1))]$$

resulting in

$$x(t_1) = x(t_0) + \frac{h}{2}[f(t_0, x(t_0)) + f(t_1, x(t_1))]$$

Since the unknown $x(t_1)$ appears on both sides of the equation, generally in a nonlinear manner, Heun applied a procedure called predictor/corrector method to find $x(t_1)$. He applied Euler's method to predict $x(t_1)$.

$$\tilde{x}_1 = x_0 + f(t_0, x_0)h$$

Substitution of this approximation of $x(t_1)$ by its predicted value $\tilde{x}_1$ in the above integral equation results in a corrected version of x_1, which is Heun's solution for t_1 node.

$$x_1 = x_0 + \frac{h}{2}[f(t_0, x(t_0)) + f(t_1, \tilde{x}_1)]$$

for other t_j, $j = 1, 2, \ldots, J$, Heun's method finds x_j by iteration. That is

$$\tilde{x}_{j+1} = x_j + hf(t_j, x_j); \quad \text{predictor of } x_{j+1}$$

$$x_{j+1} = x_j + \frac{h}{2}[f(t_j, x_j) + f(t_{j+1}, \tilde{x}_{j+1})]; \quad \text{corrector of } x_{j+1}$$

The notation convention used in this book, is to denote predicted value of $x(t_j)$ as $\tilde{x}_j$ and the corrected value computed for $x(t_j)$ as x_j.

The errors in Heun's method are due to truncation errors and approximating $f(t_{j+1}, x(t_{j+1}))$ by $f(t_{j+1}, \tilde{x}_{j+1})$. Assuming $f(t, x)$ satisfies Lipschitz condition one finds

$$| E[f]| = \frac{|f''(\xi_1)|}{12}h^3 + \frac{h}{2}|f(t_{j+1}, x(t_{j+1})) - f(t_{j+1}, \tilde{x}_{j+1})|$$

$$\leq \frac{|f''(\xi_1)|}{12}h^3 + \frac{h}{2}L|x(t_{j+1})) - \tilde{x}_{j+1}| = O(h^3)$$

Thus, Heun's method has a local error of order $O(h^3)$ for each step, and global error $O(h^2)$, which makes it more accurate than Euler's method by a factor of h. To give an example, again take IVP $x' = 1 + x^2$ with $x(0) = 0$, numerical solutions via Euler and Heun's methods are shown in Fig. 6.6.

MATLAB code for Fig. 6.6

```
% inputting initial information
xx=1.4; N=50; h= xx/N; y(1)=0.00; yh(1)=0.0;
% finding exact, Euler and Heun solutions
for m=1:N
x(m)= h*(m-1); ye(m)= tan(x(m)); y(m+1)= y(m)+ h*(1+
(y(m))^2);
p(m+1)= yh(m)+ h*(1+ (yh(m))^2);
yh(m+1)= yh(m)+ h*(1+ (yh(m))^2 + 1+ (p(m+1))^2)/2; end;
x(N+1)=xx; ye(N+1)=tan (x(N+1)); ee=y-ye; eh=yh-ye;
plot(x,ee,'k.', x,eh, 'k–','LineWidth',1.5)
```

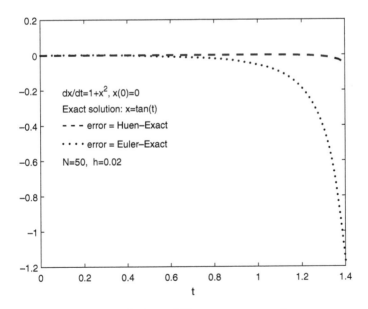

Fig. 6.6. Euler and Heun error comparison.

As expected, Fig. 6.6 shows that Heun's method more accurately reproduces the actual solution than Euler's method. It also follows that Euler's global error is $O(h)$, but Heun's global error is $O(h^2)$, consistent with results shown in Fig. 6.6.

6.4 Multistep and Predictor–Corrector Methods

We saw Heun's method produced more accurate results than Euler's method, although it had to start with Euler's results to predict a solution and correct it to compute a more accurate solution. Thus, it is natural to see if one can extend Heun's ideas and enhance the accuracy of numerical solutions. The integral in Heun's procedure

$$x(t_{i+1}) = x(t_i) + \int_{t_i}^{t_{i+1}} f(s, x(s))ds$$

was evaluated via trapezoidal quadrature procedure. In other words $f(t, x(t))$ was approximated by a straight line segment over $[t_i, t_{i+1}]$ for $t_i = a + ih$, where $h = \frac{b-a}{N}$, & $i = 0, 1, 2, \ldots, N$. Thus it is natural to ask what if one approximates $f(t, x(t))$ by a higher order polynomials. Pursuit of this idea, leads to Adams–Bashforth method, which approximates $f(t, x(t))$ by a Lagrange polynomial $\tilde{P}_{n-1}$ that interpolates nodal points (t_{i-k}, f_{i-k}), where $k = 0, 1, 2, \ldots, (n-1)$, and f_{i-k} denotes the numerical value of $f(t_{i-k}, x(t_{i-k}))$. In other words, Adams–Bashforth method approximates $f(t, x)$ by

$$f(t, x(t)) = \tilde{P}_{n-1}(t) + \tilde{R}_{n-1}(t)$$

$$\tilde{P}_{n-1}(t) = \sum_{k=0}^{n-1} f_{i-k} \, L_{n-1,k}(t), \quad \text{where}$$

$$L_{n-1,k}(t) = \prod_{\substack{j=0 \\ j \neq k}}^{n-1} \frac{(t - t_{i-j})}{(t_{i-k} - t_{i-j})} \quad \text{and}$$

$$\tilde{R}_{n-1}(x) = \frac{(t - t_i)(t - t_{i-1})(t - t_{i-2})\ldots(t - t_{i+1-n})}{n!} f^{(n)}(\xi, x(\xi))$$

Using above defined $\tilde{P}_{n-1}(t)$, one predicts value of $x(t)$ at $t = t_{i+1}$

$$\text{Predictor equation:} \quad \tilde{x}_{i+1} = x_i + \int_i^{i+1} \tilde{P}_{n-1}(s)ds$$

Similar to the Heun's procedure, Adams–Moulton developed a corrector procedure for Adams–Bashforth method, by defining another Lagrange polynomial P_n for nodal points (t_{i+1-k}, f_{i+1-k}), $k = 0, 1, 2, \ldots, n$. Since for t_{i+1}, value of x_{i+1} is not known, predicted nodal point $\tilde{x}_{i+1}$ is used to compute $f_{i+1} = f(t_{i+1}, \tilde{x}_{i+1})$. This new polynomial takes the form

$$P_n(t) = \sum_{k=0}^{n} f_{i+1-k}\, L_{n-1,k}(t), \quad L_{n-1,k}(t) = \prod_{\substack{j=0 \\ j \neq k}}^{n} \frac{(t - t_{i+1-j})}{(t_{i+1-k} - t_{i+1-j})}$$

Similar to the Heun's method, integrating $P_n(t)$ leads to Adams–Moulton procedure, that finds corrected value of x_{i+1} from its predicted value $\tilde{x}_{i+1}$.

$$\text{Corrector equation:} \quad x_{i+1} = x_i + \int_i^{i+1} P_n(s)ds$$

It is evident that such a procedure requires knowledge of the solution at $x(t_0), x(t_1), \ldots, x(t_{n-1})$. However, only the initial value $x(t_0) = x_0$ is known. Thus, to start the procedure, one needs to apply a different method for finding the starting values of the solution. A method that needs more than the initial value to start, is called multistep method. To demonstrate the procedure, let us first present the Adams–Bashforth two-step procedure.

Two-step Adams–Bashforth method

Since $n = 2$, we interpolation $f(t, x(t))$ by a Lagrange polynomial of at most degree one

$$f(t, x(t)) = \tilde{P}_1(t) + \tilde{R}_1(t)$$

where $\tilde{P}_1(t) = f_i L_{1,0}(t) + f_{i-1} L_{1,1}(t) = f_i \frac{t - t_{i-1}}{t_i - t_{i-1}} + f_{i-1} \frac{t - t_i}{t_{i-1} - t_i}$

and $\tilde{R}_1(t) = \dfrac{(t - t_i)(t - t_{i-1})f''(\xi, x(\xi))}{2}$

Following the procedure presented above the predictor equation takes
the form:

$$\tilde{x}_{i+1} = x_i + \int_{t_i}^{t_{i+1}} \left[f_i \frac{t - t_{i-1}}{t_i - t_{i-1}} + f_{i-1} \frac{t - t_i}{t_{i-1} - t_i} \right] dt$$

changing variable to $u = t - t_i$

$$\tilde{x}_{i+1} = x_i + \int_0^h \left[f_i \frac{u + h}{h} - f_{i-1} \frac{u}{h} \right] du = x_i + \frac{h}{2} [3 f_i - f_{i-1}]$$

Local error $\tilde{E}_1 = \int_{t_i}^{t_{i+1}} \tilde{R}_1(t) dt$

$$= \int_{t_i}^{t_{i+1}} \frac{(t - t_i)(t - t_{i-1}) f''(\xi, x(\xi))}{2} dt = \int_0^h \frac{u(u + h) x'''(\xi)}{2} du$$

$$\tilde{E}_1 = \frac{5h^3}{12} x'''(\xi) = O(h^3)$$

Applying same procedure for $n = 4$, the four-step Adams–Bashforth
method follows

$$\tilde{x}_{i+1} = x_i + \frac{h}{24} [55 f_i - 59 f_{i-1} + 37 f_{i-2} - 9 f_{i-3}]$$

with local error

$$\tilde{E}_3 = \frac{251 h^5}{720} x^{(5)}(\xi) = O(h^5).$$

Since extrapolation was used to find $\tilde{x}_{i+1}$, we apply the corrector
procedure developed by Adams and Moulton. To demonstrate the
procedure let us present the two-step Adams–Moulton method.

Two-step Adams–Moulton method

In this method $f(t, x(t))$ is represented by

$$f(t, x(t)) = P_n(t) + R_n(t)$$

where

$$P_n(t) = \sum_{k=0}^n f(t_{i+1-k}, x(t_{i+1-k})) L_{n,i+1-k}(t) \ \&$$

$$L_{n,j} = \prod_{\substack{k=0 \\ j \neq (i+1-k)}}^n \frac{(t - t_{i+1-k})}{(t_j - t_{i+1-k})}$$

$$\text{and} \quad R_n(t) = \frac{f^{(n+1)}(\xi, x(\xi))}{(n+1)!} \prod_{k=0}^{n}(t - t_{i+1-k})$$

For $n = 2$, we have

$$P_n(t) = f(t_{i+1}, x_{i+1})\alpha(t) + f(t_i, x_i)\beta(t) + f(t_{i-1}, x_{i-1})\gamma(t)$$

$$R_2(t) = \frac{f'''(\xi, x(\xi))}{6}(t - t_{i+1})(t - t_i)(t - t_{i-1})$$

where

$$\alpha(t) = \frac{(t - t_i)(t - t_{i-1})}{(t_{i+1} - t_i)(t_{i+1} - t_{i-1})}, \quad \beta(t) = \frac{(t - t_{i+1})(t - t_{i-1})}{(t_i - t_{i+1})(t_i - t_{i-1})}$$

$$\text{and} \quad \gamma(t) = \frac{(t - t_{i+1})(t - t_i)}{(t_{i-1} - t_{i+1})(t_{i-1} - t_i)}$$

Change variable $t = u + t_i$ and Integrating, leads to

$$\int_{t_i}^{t_{i+1}} \alpha(t)dt = \int_0^h \frac{u(u+h)}{h(2h)}du = \left.\frac{u^3/3 + hu^2/2}{2h^2}\right|_0^h = \frac{5h}{12}$$

Likewise

$$\int_{t_i}^{t_{i+1}} \beta(t)dt = \frac{2h}{3}, \quad \int_{t_i}^{t_{i+1}} \gamma(t)dt = -\frac{h}{12} \quad \text{and local error}$$

$$E_2 = \int_{t_i}^{t_{i+1}} R_2(t)dt = -\frac{h^4}{24}x^{(4)}(\xi).$$

It follows that the two-step Adams–Moulton method has the form

$$x_{i+1} = x_i + \frac{h}{12}[5f(t_{i+1}, \tilde{x}_{i+1}) + 8f(t_i, x_i) - f(t_{i-1}, x_{i-1})]$$

Similar procedure is used to develop a three-step Adams–Moulton method

$$x_{i+1} = x_i + \frac{h}{24}[9f(t_{i+1}, \tilde{x}_{i+1}) + 19f(t_i, x_i) - 5f(t_{i-1}, x_{i-1})$$
$$+ f(t_{i-2}, x_{i-2})]$$

with local truncation error $E_3 = -\frac{19h^5}{720}x^{(5)}(\xi) = O(h^5)$.

Combining Adams–Bashforth and Adams–Moulton procedures results in a predictor corrector procedure referred to as Adams–Bashforth–Moulton method. The four-step Adams–Bashforth predictor and three-step Adams–Moulton corrector combination has seen extensive use. The error for such a procedure can be estimated by noting for four-step Adams–Bashforth procedure. the local error is

$$\tilde{E}_3 = x(t_{i+1}) - \tilde{x}_{i+1} = \frac{251}{720} x^{(5)}(\xi_1) h^5$$

and for three-step Adams–Moulton procedure the local error is

$$E_3 = x(t_{i+1}) - x_{i+1} = -\frac{19}{720} x^{(5)}(\xi_2) h^5$$

To make the right-hand side of above equations the same, multiply them by appropriate constants and add the resulting equations to find

$$(19 + 251) x(t_{i+1}) - 19\tilde{x}_{i+1} - 251 x_{i+1}$$
$$= \frac{19 * 251}{720} [x^{(5)}(\xi_1) - x^{(5)}(\xi_2)] h^5$$

Assuming $x^{(5)}(\xi_1) \approx x^{(5)}(\xi_2)$ and adding $\pm 19 x_{i+1}$ to the above equation,

$$270 x(t_{i+1}) - 270 x_{i+1} + 19 x_{i+1} - 19\tilde{x}_{i+1} \approx 0,$$
$$\text{or } x(t_{i+1}) - x_{i+1} \approx -\tfrac{19}{270}(x_{i+1} - \tilde{x}_{i+1})$$

Since predicted $\tilde{x}_{i+1}$ and corrected x_{i+1} values of $x(t_{i+1})$ are numerically known, the above equation gives an estimate of error made in finding the solution via Adams–Bashforth–Moulton procedure.

$$\frac{270}{720} x^{(5)}(\xi_2) h^5 \approx x_{i+1} - \tilde{x}_{i+1}$$

The above relation allows us to estimate nodal spacing h needed for a desired tolerance level, in a similar manner as developed for RKF45.

Above derivation shows Adams–Bashforth–Moulton method error is $O(h^5)$ for each step, and one needs x_0, x_1, x_2, x_3 associated with t_0, t_1, t_2, t_3 to start the procedure. (t_0, x_0) is the initial condition which is given, but for the other needed starting points, one may

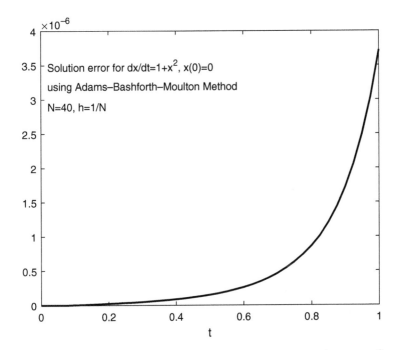

Fig. 6.7. Adams–Bashforth–Moulton solution error for $\frac{dx}{dt} = 1 + x^2$.

apply RK4 method to find them. A numerical example is presented in Fig. 6.7.

MATLAB code for Fig. 6.7

```
% inputting initial information
N=40; h= 1/N; hh=h/2; yb(1)=0;
% finding exact solution
for m=1:N
t(m)= h*(m-1); ye(m)= tan(t(m)); end;
t(N+1)=1; ye(N+1)=tan( t(N+1));
% applying RK4 to find starting values for m=1:3
f1= 1+(yb(m))²; f2=1+(yb(m)+hh*f1)²; f3=1+(yb(m)+hh*f2)²;
f4=1+(yb(m)+h*f3)²; yb(m+1)= yb(m)+ h*(f1+2*f2+2*f3+f4)/6;
end;
% finding Adams–Bashforth–Moulton solution
for m=4:N
fm3=1+ (yb(m-3))²; fm2=1+ (yb(m-2))²; fm1=1+ (yb(m-1))²;
```

fm0=1+ (yb(m))²;
pk1=yb(m)+(h/24)*(-9*fm3+37*fm2-59*fm1+55*fm0);
yb(m+1)=yb(m)+ (h/24)*(fm2-5*fm1+19* fm0+ 9*(1+(pk1)²));
end;
% finding the difference
dabm=yb-ye; plot (t, dabm, 'k-','LineWidth', 1.3)

If one compares results shown in Fig. 6.4 that showed RK4 error performance to Fig. 6.7 that shows Adams–Bashforth–Moulton error performance, the figures indicate RK4 performed better than Adams–Bashforth–Moulton predictor–corrector method for the example presented. However, it has been pointed out that in order to have a fair comparison when comparing two different procedures, one needs to adjust step sizes in order to make number of operations in the procedures to be comparable.

There are other multistep and predictor-corrector methods. For example Milne–Simpson method, where Milne method provides the predictor solution, and like Heun's method that used trapezoidal rule, Milne–Simpson method uses Simpson quadrature to find the corrector solution.

$$\text{Predictor:} \quad \tilde{x}_{i+1} = x_{i-3} + \frac{4h}{3}[2f_i - f_{i-1} + 2f_{i-2}],$$

$$i = 3, 4, \ldots (N-1)$$

with local error of $\frac{14h^5}{45}x^{(5)}(\xi_i)$, and

$$\text{Corrector:} \quad x_{i+1} = x_{i-1} + \frac{h}{3}[\tilde{f}_{i+1} + 4f_i + f_{i-1}],$$

$$i = 1, 2, \ldots, (N-1)$$

with local error of $-\frac{h^5}{90}x^{(5)}(\xi_i)$, where $\tilde{f}_{i+1} \equiv f(t_{i+1}, \tilde{x}_{i+1})$.

Comparing Adams–Bashforth–Moulton local errors with Milne–Simpson method, one notes Milne–Simpson local errors are smaller, however due to stability issues with Milne–Simpson, the method has seen limited use. This observation motivates an introduction to the basics of numerical stability for multistep methods.

6.5 Stability for General Linear Multistep Methods

In this section, we present stability conditions for general linear m-step equation of the form

$$\frac{x_{j+1} - \alpha_1 x_j - \alpha_2 x_{j-1} - \alpha_3 x_{j+1-3} \cdots - \alpha_m x_{j+1-m}}{h}$$

$$= \beta_0 f_{j+1} + \beta_1 f_j + \beta_2 f_{j-1} + \beta_3 f_{j+1-3} \cdots - \beta_m f_{j+1-m}$$

It is assumed local errors for starting values $x_1, x_2, \ldots, x_{m-1}$ which are found using other methods are sufficiently small and local truncation error for above m-step equation is at least $O(h)$ as $h \to 0$. To motivate derivation of stability condition let us initially assume $f(t, x) = 0$ and define the standard characteristic polynomial $P(\lambda)$ for this simplified linear multistep equation

$$x_{j+1} - \alpha_1 x_j - \alpha_2 x_{j-1} - \alpha_3 x_{j+1-3} \cdots - \alpha_m x_{j+1-m} = 0$$
$$P(\lambda) = \lambda^m - a_1 \lambda^{m-1} - a_2 \lambda^{m-2} \cdots - a_{m-1} \lambda - a_m$$

Since numerical stability study is interested in seeing if the error grows, let us define $y_j = x(t_j) - x_j$ and note that

$$y_{j+1} - \alpha_1 y_j - \alpha_2 y_{j-1} - \alpha_3 y_{j+1-3} \cdots - \alpha_m y_{j+1-m} = 0$$

From the above, it follows that if λ is a zero of $P(\lambda)$, then $y_j = \lambda^j$ is the solution of

$$y_{j+1} - \alpha_1 y_j - \alpha_2 y_{j-1} - \alpha_3 y_{j+1-3} \cdots - \alpha_m y_{j+1-m}$$
$$= \lambda^{j+1} - \alpha_1 \lambda^j - \alpha_2 \lambda^{j-1} - \alpha_3 \lambda^{j+1-3} \cdots - \alpha_m \lambda^{j+1-m}$$
$$= \lambda^{j+1-m} [\lambda^m - \alpha_1 \lambda^{m-1} - \alpha_2 \lambda^{m-2} \cdots - \alpha_m] = \lambda^{j+1-m} P(\lambda) = 0$$

Since error $y_j = \lambda^j$, then if $|\lambda| > 1$ the errors grows as j increases and the solution does not remain numerically stable. This finding leads to the followings.

Definition of root condition

Let λ_i be the roots of characteristic polynomial $P(\lambda)$ for a linear multistep procedure. Then root condition is said to be satisfied if

1. $|\lambda_i| \leq 1$ for all λ_i being roots of $P(\lambda)$, and
2. If $|\lambda_i| = 1$ is a root of $P(\lambda)$, then λ_i needs to be a simple root.

From the above, it follows that when $f(t, x)$ is identically zero, then a linear multistep method is stable if $P(\lambda)$ satisfies the root condition. It has been shown that the same stability result also applies when $f(t, x)$ is not identically zero (Isaacson and Keller, 1966), leading to the following theorem.

Theorem. *A linear multistep method is stable if and only if the root condition is satisfied.*

Example 1. Adams–Bashforth and Adams–Moulton methods

We notice for both Adams–Bashforth four-step method ($m = 4$) and Adams–Moulton three-step method ($m = 3$) their characteristic polynomials take the form

$$P(\lambda) = \lambda^m - \lambda^{m-1} = \lambda^{m-1}(\lambda - 1) = 0$$

Thus, $\lambda = 0$ with multiplicity $(m-1)$ and $\lambda = 1$, a simple root, are the only roots of characteristic polynomials $P(\lambda)$ for Adams–Bashforth and Adams–Moulton methods. Thus, these methods satisfy the root condition, and are stable. This stability property has made application of four-step Adams Bashforth predictor along with three-step Adams Moulton corrector method to be commonly used, and for the sake of simplicity of the presentation, this combination is labeled as Adams–Bashforth–Moulton procedure.

Definition of strongly and weakly stable methods

If the characteristic polynomial $P(\lambda)$ of a linearly stable multistep method has $\lambda = 1$ as the only root satisfying $|\lambda| = 1$ condition, then the method is said to be "Strongly Stable," otherwise it is said to be "Weakly Stable." It follows that Adams–Bashforth–Moulton method is strongly stable.

Example 2. Milne method

We saw the Milne equation has the form

$$\frac{x_{j+1} - x_{j-3}}{h} = \frac{x_{j+1} - \alpha_4 x_{j-3}}{h} = \beta_1 f_j + \beta_2 f_{j-1} + \beta_3 f_{j-2}$$

Hence, $P(\lambda) = \lambda^m - a_1\lambda^{m-1}\ldots - a_{m-1}\lambda - a_m$, with $m = 4$ and $\alpha_1 = \alpha_2 = \alpha_3 = 0, \alpha_4 = 1$. Thus

$$P(\lambda) = \lambda^4 - a_4\lambda^{4-4} = \lambda^4 - 1 = 0$$

with its roots being

$$\lambda = e^{\frac{i2n\pi}{4}}, \quad \text{for } n = 0, 1, 2, 3.$$

In the above formula, the symbol i stands for the imaginary number $i = \sqrt{-1}$. Since Milne characteristic polynomial has four roots

$$\lambda_0 = 1, \lambda_1 = i, \ \lambda_2 = -1, \ \& \ \lambda_3 = -i$$

and all the four roots of $P(\lambda)$ have unit magnitude, $|\lambda| = 1$, with $\lambda = 1$ not being the only root satisfying $|\lambda| = 1$, it follows from the above that Milne method is weakly stable.

To provide a numerical example for a weakly stable method let us find solution to the following ODE

$$\frac{dx}{dt} = \gamma x, \text{ with } x(0) = 1 \ \& \ \gamma = -3$$

by applying the Milne method

$$x_{j+1} = x_{j-3} + \frac{4h}{3}[2f_j - f_{j-1} + 2f_{j-2}]$$

The Milne solution for this example is shown in Fig. 6.8.

MATLAB code for Fig. 6.8

```
% inputting initial information
a=0; b=3; N=20; h=(b-a)/N; r=-3; hh=h/2;y(1)=1;
% applying RK4 to find starting values
for m=1:3 f1= r*y(m); f2=r*(y(m)+hh*f1); f3=r*(y(m)+hh*f2);
f4=r*(y(m)+h*f3); y(m+1)= y(m)+ h*(f1+2*f2+2*f3+f4)/6; end;
% finding exact solution
for n=1:N x(n)=a+(n-1)*h; ye(n)=exp(r*x(n)); end
% finding Milne solution
for  n=4:N  y(n+1)=y(n-3)+(4*h*r/3)*(2*y(n)-y(n-1)+2*y(n-2));
end for n=1:N z(n)=y(n); end; plot(x,ye,'k-', x, z, 'k-','LineWidth',
1.3)
```

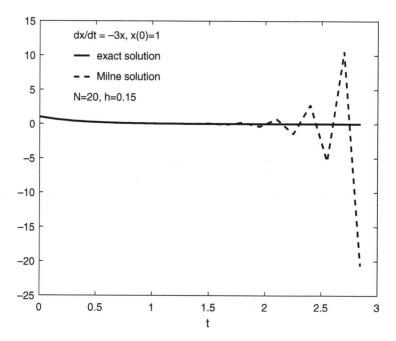

Fig. 6.8. Solution to $\frac{dx}{dt} = -3x, x(0) = 1$, via Milne procedure.

Figure 6.8 shows for $\gamma = -3$, as time increases the solution starts to oscillate around the exact solution. This oscillation continues to be present even if h is reduced to smaller values.

In order to show the oscillation is not due to the differential equation considered, but is due to the method used to find the solution to $x' = -3x$, let us try to solve the same problem using RK4 method. The numerical result using RK4 procedure is shown in Fig. 6.9.

MATLAB code for Fig. 6.9

```
% inputting initial information
a=0; b=3; N=20; h=(b-a)/N; r=-3; hh=h/2;y(1)=1;
% applying RK4 to find solution to x'=-3x
% finding exact solution
for n=1:N
x(n)=a+(n-1)*h;
ye(n)=exp(r*x(n));
end
```

% finding RK4 solution
for m=1:N
f1= r*y(m);
f2=r*(y(m)+hh*f1); f3=r*(y(m)+hh*f2);
f4=r*(y(m)+h*f3);
y(m+1)= y(m)+ h*(f1+2*f2+2*f3+f4)/6; end
for n=1:N
z(n)=y(n); end
plot(x,ye,'k-', x, z, 'k–','LineWidth', 1.3)

As it can be seen from Fig. 6.9, for $\gamma = -3$, as time increases the solution found numerically via RK4 method unlike the Milne method does not oscillate and it continues to well represent the exact solution.

It is also interesting to note if we change the coefficient γ in the differential equation $x' = \gamma x$ from $\gamma = -3$ to $\gamma = 3$ the oscillation about exact solution does not show up and Milne method also provides a satisfactory result as shown in Fig. 6.10. Furthermore as h

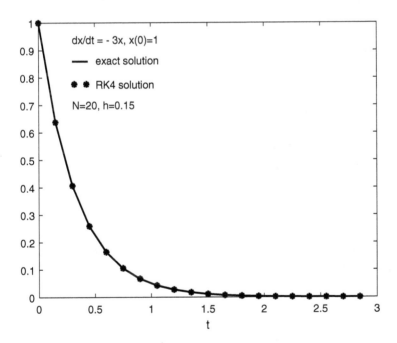

Fig. 6.9. Solution to $\frac{dx}{dt} = -3x, x(0) = 1$, via RK4 procedure.

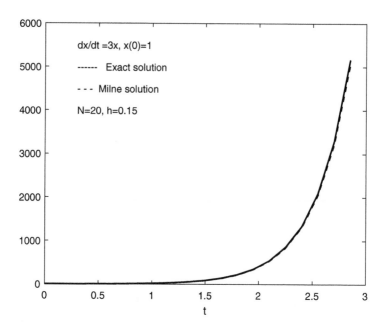

Fig. 6.10. Solution to $\frac{dx}{dt} = 3x, x(0) = 1$, via Milne procedure.

decreases the agreement between exact and numerical solution for $\gamma = 3$ becomes better.

MATLAB code for Fig. 6.10 and numerical results using Milne procedure for $x' = 3x$ are shown in the following.

MATLAB code for Fig. 6.10

```
% inputting initial information
a=0; b=3; N=20; h=(b-a)/N; r=3; hh=h/2;y(1)=1;
% applying RK4 to find starting values
for m=1:3 f1= r*y(m); f2=r*(y(m)+hh*f1); f3=r*(y(m)+hh*f2);
f4=r*(y(m)+h*f3); y(m+1)= y(m)+ h*(f1+2*f2+2*f3+f4)/6; end;
% finding exact solution
for n=1:N x(n)=a+(n-1)*h; ye(n)=exp(r*x(n)); end
% finding Milne solution
for   n=4:N   y(n+1)=y(n-3)+(4*h*r/3)*(2*y(n)-y(n-1)+2*y(n-2));
end for n=1:N z(n)=y(n); end; plot(x,ye,'k-', x, z, 'k-','LineWidth',
1.3)
```

6.6 Higher-Order ODE and Initial Value Problems

We presented IVP methods for first order differential equations. Next, we study how to apply these methods to higher order ordinary differential equations. We saw higher order equations can be written as a system of a first order equations. For example, Newton's equation

$$mx'' + cx' + kx = f(t)$$

can be written as

$$X = \begin{bmatrix} x \\ x' \end{bmatrix} = \begin{bmatrix} x_1 \\ x_2 \end{bmatrix}$$

and

$$X' = \begin{bmatrix} x' \\ x'' \end{bmatrix} = \begin{bmatrix} x' \\ -\dfrac{c}{m}x' - \dfrac{k}{m}x + \dfrac{f(t)}{m} \end{bmatrix} = \begin{bmatrix} F_1(t, x_1, x_2) \\ F_2(t, x_1, x_2) \end{bmatrix} = F(t, X)$$

or

$$X' = \begin{bmatrix} 0 & 1 \\ -\dfrac{k}{m} & -\dfrac{c}{m} \end{bmatrix} \begin{bmatrix} x_1 \\ x_2 \end{bmatrix} + \begin{bmatrix} 0 \\ \dfrac{f(t)}{m} \end{bmatrix} = F(t, X)$$

with the initial condition $X(t_0) = \begin{bmatrix} x(t_0) \\ x'(t_0) \end{bmatrix} = X_0$.

The ability to reduce higher-order differential equations to a system of first-order equations allows procedures developed for scalar initial value problems be applied to its vector version too. For example, if we like find solution of the above Newton equation using Heun's method, we first use Euler's method to predict the solution and then use trapezoidal quadrature to find its correction.

$$\tilde{X}_{k+1} = X_k + hF(t_k, X_k), \quad k = 0, 1, 2, 3, \ldots, N$$

$$X_{k+1} = X_k + \frac{h}{2}[F(t_k, X_k) + F(t_{k+1}, \tilde{X}_{k+1})]$$

Similar procedure can be applied when using RK4 or multi-step procedures. Of course, the computational work will be more tedious and will consume more computer time. To give an example and avoid too many indices let us consider following second-order initial value problem

$$x'' + x = 0, \quad \text{with} \quad x(0) = 1 \ \& \ x'(0) = 0$$

Let's call $v = x'$, and define $X = \begin{bmatrix} x \\ v \end{bmatrix}$. It follows

$$X' = \begin{bmatrix} x' \\ v' \end{bmatrix} = \begin{bmatrix} v \\ -x \end{bmatrix} = F(t, X)$$

Defining numerical approximation of $x(t_k)$ and $v(t_k)$ as x_k and v_k, respectively, leads to

$$X_k = \begin{bmatrix} x_k \\ v_k \end{bmatrix}, \quad \text{with} \quad X_0 = \begin{bmatrix} x_0 \\ v_0 \end{bmatrix} = \begin{bmatrix} x(0) \\ x'(0) \end{bmatrix} = \begin{bmatrix} 1 \\ 0 \end{bmatrix}.$$

Applying for example Euler's method leads to

$$X_{k+1} = X_k + hF(t_k, X_k), \quad k = 0, 1, 2, \ldots, N$$

Since initial value X_0 is known, solution X_k can be found recursively from above equation.

The condition for uniqueness of solution for a system of equations is again Lipschitz condition, where absolute value for scalar case is now replaced by a chosen vector norm and its associated matrix norm. That is, $F(t, X)$ is Lipschitz if exist a finite scalar L such that

$$\|F(t, X) - F(t, Y)\| \leq L\|X - Y\|$$

for all $t, X \ \& \ Y$ in the domain of their definition. For the example considered, $x'' = -x$, $x(0) = 1$, $x'(0) = 0$, written as a first-order vector equation

$$X' = \begin{bmatrix} x' \\ v' \end{bmatrix} = \begin{bmatrix} v \\ -x \end{bmatrix} = F(t, X)$$

The Lipschitz condition takes the form

$$\|F(t, X) - F(t, Y)\|$$

$$= \left\| \begin{bmatrix} v \\ -x \end{bmatrix} - \begin{bmatrix} w \\ -y \end{bmatrix} \right\| = \left\| \begin{bmatrix} 0 & 1 \\ -1 & 0 \end{bmatrix} (X - Y) \right\| \leq L\|X - Y\|$$

where $Y = \begin{bmatrix} y \\ w \end{bmatrix}$, $w = y'$ and $L = \left\| \begin{bmatrix} 0 & 1 \\ -1 & 0 \end{bmatrix} \right\|$. Since $L < \infty$, $F(t, X)$ is Lipschitz, therefor the solution to above initial value problem is unique. In order to see numerical performance for such a initial value for this system of equations, let us note the exact solution for the problem is

$$X(t) = \begin{bmatrix} \cos t \\ -\sin t \end{bmatrix}$$

Numerical result using Euler's method is shown in Fig. 6.11.

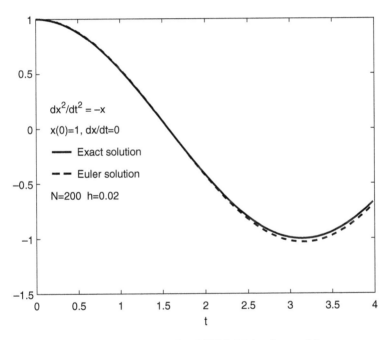

Fig. 6.11. Higher order ODE initial value problem.

MATLAB code for Fig. 6.11.

```
% inputting initial information a=0; b=4; N=200; h=(b-a)/N;y
(:,1)=[1;0];
for n=1:N
% computing exact solution
x(n)=a+(n-1)*h; ye(n)=cos(x(n)); end
```

N1=N-1;
% finding Euler's solution for higher order ODE
for n=1:N1
y(:,n+1)=y(:,n)+ h* [0 1;-1 0]*y(:,n); end
yn=y(1,:); plot(x,ye,'k-', x, yn, 'k–','LineWidth', 1.3)

Exercises

6.1. Let

$$\frac{dx}{dt} = f(t, x) = t(x - 1)^{\frac{1}{5}}$$

satisfying the initial condition $x(t_0) = x_0$.

(a) For what values of x and t, $f(t, x)$ satisfies Lipschitz condition in x?

(b) For what initial value of x_0 the solution to above initial value problem is unique?

6.2. (a) Prove that a solution to the integral equation

$$x(t) = x(t_0) + \int_{t_0}^{t} f(s, x(s))ds$$

is a solution of $x' = f(t, x)$ with the initial value $x(t_0) = x_0$.

(b) Picard's method finds the solution to the initial value problem, via solving above integral equation by following iteration method.

$$x_{n+1}(t) = x(t_0) + \int_{t_0}^{t} f(s, x_n(s))ds$$

(c) For the initial value problem $x' = (t - x)^2$, with $t_0 = 1$ & $x(t_0) = 2$, find $x(t)$ via the Picard's method. Find only $x_1(t)$ & $x_2(t)$.

6.3. Apply Euler's method to approximately find solution to the initial value problem

$$x' + 2tx = 2t^2 + 1, \quad \text{with } x(0) = 1, \quad \text{for } 0 \leq t \leq 2$$

choose h such that the global error is not larger than 0.01.

Compare your result with the exact solution $x = e^{-t^2} + t$ and see if the error is within the expectation.

6.4. Consider IVP

$$x' = t^{-1}(x-1)^{\frac{2}{3}}, \quad \text{with } x(1) = 2 \ \& \ 1 \le t \le 2$$

(a) Is the solution to this initial value problem unique? Justify your answer.

(b) Apply Euler's method to approximately find a solution to this IVP. Choose $h = 0.1$.

(c) Estimate the global error of the found solution at $t = 2$.

6.5. Apply RK4 method to find the approximate solution to the initial value problem

$$x' + x = (t+1)^2, \quad \text{with } x(0) = 2, \quad \text{for } 0 \le t \le 2$$

Use $N = 20$.

Estimate the error and compare it to the exact solution $x = e^{-t} + t^2 + 1$.

6.6. Same as Problem 6.3, except use Heun's method to find the solution.

6.7. (a) Same as Problem 6.4, except use Heun's method to find the solution.

(b) Find the exact solution to Problem 6.4.

(c) Compare the estimated global error for the Heun's method and the difference between exact and Heun's approximate solution at $t = 2$.

6.8. Same as Problem 6.5, except use Adams–Bashforth–Moulton method to find the solution.

6.9. (a) Show that Euler's method satisfies the root conditions.

(b) Is Euler's method a weakly or strongly stable procedure? Justify your answer.

(c) Applying the central difference approximation

$$x' \approx \frac{x_{n+1} - x_{n-1}}{2h}$$

to $x' = f(t,x)$ leads to the following finite difference equation

$$\frac{x_{n+1} - x_{n-1}}{2h} = f(t, x_n)$$

If this approximation is used to find solution to the initial value problem $x' = f(t,x)$ with $x(t_0) = x_0$, what is the estimated local error.

(d) Does the equation in part (c) satisfies the root conditions? If yes, is it a weakly or strongly stable method?

6.10. Find the approximate solution to

$$x' = \lambda x, \quad \text{with } x(0) = 1, \quad \text{for } 0 \le t \le 3$$

by applying following discretization:

$$\frac{x_{n+1} - x_{n-1}}{2h} = \lambda x_n$$

to arrive at an iterative procedure to find an approximate solution. Take $N = 30$ and to start the procedure use Euler's method to find x_1.

(a) Take $\lambda = -2$ and compare your approximate solution to its exact solution.
(b) Same as part (a) except choose $\lambda = 2$.
(c) Same as part (a), except use Heun's method to find its approximate solution. Compare it with solution found in part(a).
(d) Same as part (b), except use Heun's method to find its approximate solution. Compare it with the solution found in part(b).

6.11. Find the approximate solution of the following initial value problem:

$$x'' + 2x' - 8x = 0, \quad \text{for } 0 \le t \le 2$$

satisfying the initial condition

$$x(0) = 3 \ \& \ x'(0) = 0$$

Hint: Reduce this second-order ODE by writing it as a system of first-order ODE, $\mathbf{X}' = \mathbf{F}(t, \mathbf{X})$.

(a) Apply the Euler's method for $N = 100$ to find an approximate solution.
(b) Apply the Heun's methods for $N = 100$ to find an approximate solution.
(c) Compare your results to the exact solution of this IVP.

6.12. Consider the initial value problem

$$x'' + 2x' - 8x = 0, \quad \text{for } 0 \le t \le 2$$

satisfying the initial conditions

$$x(0) = 3 \ \& \ x'(0) = 0$$

(a) Write this second order equation as a system of first order differential equation:

$$\mathbf{X}' = \mathbf{F}(t, \mathbf{X}), \text{ with } \mathbf{X}(0) = \begin{bmatrix} 3 \\ 0 \end{bmatrix}$$

(b) Prove that $\mathbf{F}(t, \mathbf{X})$ satisfies Lipschitz condition.
(c) Apply RK4 method with $N = 20$ to find the approximate solution to this initial value problem.
(d) Compare RK4 solution to the exact solution to find its error.

6.13. Consider the initial value problem:

$$x''' + 2tx'' + 3x' - 2tx = -t^3 + 5t, \quad \text{for } t \in [0, 2]$$

satisfying initial conditions

$$x(0) = 1, \ x'(0) = 0 \ \& \ x''(0) = -1.$$

(a) Find an approximate solution to this IVP using the Euler's method for $N = 20$ and $t \in [0, 2]$.
(b) Apply RK4 method with $N = 20$ to find the approximate solution to this initial value problem for $t \in [0, 2]$.
(c) The exact solution to this IVP is

$$x(t) = e^{-t^2} + \frac{1}{2}t^2$$

Compare your found solutions via Euler's method and RK4 procedure. Discuss their performance and the associated errors.

6.14. Same as Problem 6.13 except use Adams–Bashforth–Moulton predictor corrector method to find an approximate solution to the given IVP. Apply RK4 method to start the Adams–Bashforth–Moulton procedure and estimate its global error.

6.15. Consider the initial value problem

$$x''' - 2xx'' + x'^2 = 0$$

with initial conditions

$$x(0) = x'(0) = 0 \ \& \ x''(0) = 1, \quad \text{for } t \in [0, 2]$$

(a) Write the equation as a system of first order ODE, $\mathbf{X}' = \mathbf{F}(t, \mathbf{X})$.
(b) Find the Domain (if there is any) where $\mathbf{F}(t, \mathbf{X})$ is Lipschitz.
(c) Find an approximate solution to this initial value problem using Adams–Bashforth–Moulton predictor corrector method, for $N = 20$. Apply RK2 method to start the Adams–Bashforth–Moulton procedure.

Chapter 7

Boundary Value Problems

Another topic of practical importance is the boundary value problems (BVP) for ordinary second-order differential equation. A simple example is a violin's strings where both sides of the strings are clamped, and they generate different musical notes as the violin is played. More formally stated BVP considered in this chapter takes the form

$$x'' = f(s, x, x'), \quad s \in [a, b]$$

with values of $x(s)$ and or a combination of $x(s)$ and it derivative $x' = \frac{dx}{ds}$ specified on the boundary. The most common boundary conditions appearing in applications has the form

$$x(a) = \alpha \quad \text{and} \quad x(b) = \beta$$

where α and β are some given constants. The general study of boundary value problems usually breaks down into two major cases. One being linear and the other nonlinear differential equations.

7.1 Linear Second-Order ODE Boundary Value Problems

Let us denote the linear second-order differential equations to be studied as

$$x'' = f(t, x, x') = p(s)x' + q(s)x + r(s), \text{with } x(a) = \alpha \ \& \ x(b) = \beta$$

where $x' = \frac{dx}{ds}$ and $p(s)$, $q(s)$ and $r(s)$ are continuous functions of $s \in [a, b]$. A standard method to find numerical solution for above equation is called shooting method, where one finds the solution to an initial value problem, with initial condition $x(a) = \alpha$ & $x'(a) = \gamma$, with α being given by the boundary data at $s = a$, and γ is to be chosen in such a way that $x(b) = \beta$. In other words, by adjusting γ the solution is made to satisfy $x(a) = \alpha$ and also the other boundary condition $x(b) = \beta$. This is why the method is called shooting method, since by adjusting direction of a gun, enables one to hit the target.

As we saw for solution of an initial value problem to exist, we need $f(s, x, x')$ to be continuous as a function of s, x & x' for all values of $a \leq s \leq b$ and all values of x & x' are to be in the domain of $f(s, x, x')$. Furthermore, for the solution to be unique, we needed f to satisfy Lipschitz condition, which for the case under consideration reduces to requiring $p(s)$ & $q(s)$ to be bounded for $a \leq s \leq b$.

Assuming the needed conditions for existence of the initial value problem are satisfied, the question remains how to find γ so that the initial value problem with $x(a) = \alpha$ and $x'(a) = \gamma$ will also satisfy the boundary condition $x(b) = \beta$. The standard way to find such a γ numerically is to solve two initial value problems. One being

$$y'' = p(s)y' + q(s)y + r(s), \text{ with } y(a) = \alpha \ \& \ y'(a) = 0$$

The other solution $z(s)$ is required to satisfy

$$z'' = p(s)z' + q(s)z \text{ with } z(a) = 0 \ \& \ z'(a) = 1$$

Such a set of initial conditions make y and z to be linearly independent, enabling us to represent the desired solution as

$$x(s) = y(s) + cz(s)$$

where coefficient c is a constant, that needs to be deduced from the other boundary condition. To show $x(s)$ satisfies the desired differential equation, take the second derivative of $x(s)$.

$$x''(s) = y''(s) + cz''(s) = p(s)(y(s) + cz(s))'$$
$$+ q(s)(y(s) + z(s)) + r(s)$$

Concluding that the defined $x(s)$ satisfies the differential equation of interest

$$x'' = p(s)x' + q(s)x + r(s)$$

For $x(s)$ to satisfy the desired boundary condition, we note by construction

$$x(a) = y(a) + cz(a) = \alpha \quad \text{and} \quad x(b) = y(b) + cz(b) = \beta$$

Thus, to satisfy the boundary condition at $s = b$, we need to choose coefficient

$$c = \frac{\beta - y(b)}{z(b)}$$

if $z(b) \neq 0$. We should note that if $z(b) = 0$, the given boundary value problem does not have a unique solution, since $z(a) = z(b) = 0$ and $z(s)$ can be added to any solution that satisfies the given boundary condition and the resulting solution will be a different solution satisfying the same boundary condition. However, for $z(b) \neq 0$, the above procedure leads to a unique IVP solution $x(s)$ that also satisfies boundary conditions $x(a) = \alpha$ and $x(b) = \beta$. The reader may wonder if mentioned solution by choosing an appropriate value of $\gamma = x'(a)$ is the same as solution found via setting $x = y + cz$. To show this is indeed the case, let us note that

$$\gamma = x'(a) = y'(a) + cz'(a)$$

and recall $y'(a) = 0$ and $z'(a) = 1$. Hence, $\gamma = x'(a) = 0 + c * 1 = c$. In other words, the above method is a systematic way to find $\gamma = x'(a)$ that results in a solution $x(s)$ satisfying the given boundary conditions.

In order to make the shooting method more transparent, let us consider following example

$$x'' = -4x, \quad x(0) = 2 \ \& \ x\left(\frac{\pi}{4}\right) = 1$$

To solve the problem, using the shooting method, one finds solutions for the two different initial value problems.

$$y'' = -4y, \quad y(0) = 2 \ \& \ y'(0) = 0$$

and

$$z'' = -4z, \quad z(0) = 0 \ \& \ z'(0) = 1$$

Numerical solution for y and z using any of the IVP methods discussed can be used to complete the procedure, but for sake of clarity of presentation, let us first demonstrate the shooting method analytically for this chosen example, whose analytical solutions are well known:

$$y(s) = A\cos 2s + B\sin 2s \quad \text{with } y(0) = 2,\ y'(0) = 0$$

Implementing the initial conditions: $y(0) = A + 0 = 2$, and $y'(0) = -2A\sin 0 + 2B\cos 0 = B = 0$. Thus, the first solution takes the form

$$y(s) = 2\cos 2s$$

Similarly, the second solution is also found.

$$z(s) = D\cos 2s + E\sin 2s, \text{ with } z(0) = 0 \ \& \ z'(0) = 1$$

Implying, $z(0) = D = 0$, $z'(0) = 1 = -2D\sin 0 + 2E\cos 0 = 2E = 1$, resulting in $z(s) = \frac{1}{2}\sin 2s$ and $c = \frac{\beta - y(b)}{z(b)} = \frac{1 - 2\cos\frac{2\pi}{4}}{\frac{1}{2}\sin\frac{2\pi}{4}} = 2$. Combining above results, one finds

$$x(s) = y + cz = 2\cos 2s + \sin 2s$$

To check $x(s)$ is the desired solution, let us take second derivative of $x'' = -4x$ and evaluate $x(0) = 2\cos 0 + 0 = 2$ and $x(\frac{\pi}{4}) = 2\cos(\frac{\pi}{2}) + \sin(\frac{\pi}{2}) = 1$, indicating the found solution is indeed the solution to the desired boundary value problem.

In the above example since the solution to the differential equation considered was well known, we presented an analytical approach to find its solution. Now, let us try to solve it numerically, since in most situations we don't know the solution analytically. To demonstrate how to find a BVP solution numerically, let's consider: $x'' + 4x = 0$ satisfying $x(0) = 2$ and $x(\frac{\pi}{4}) = 1$, and use Euler's method to solve the problem numerically. To do so, we first write the defined functions $y(s)$ & $z(s)$ and their derivatives as system of first-order differential equations.

$$V = \begin{bmatrix} y \\ y' \end{bmatrix} = \begin{bmatrix} v_1 \\ v_2 \end{bmatrix} \Rightarrow V' = \begin{bmatrix} v_2 \\ -4v_1 \end{bmatrix} = \begin{bmatrix} 0 & 1 \\ -4 & 0 \end{bmatrix} V = F(V)$$

with $V(0) = \begin{bmatrix} 2 \\ 0 \end{bmatrix}$, and

$$U = \begin{bmatrix} z \\ z' \end{bmatrix} = \begin{bmatrix} u_1 \\ u_2 \end{bmatrix} \Rightarrow U' = \begin{bmatrix} u_2 \\ -4u_1 \end{bmatrix} = \begin{bmatrix} 0 & 1 \\ -4 & 0 \end{bmatrix} U = G(U)$$

with $U(0) = \begin{bmatrix} 0 \\ 1 \end{bmatrix}$.

The above systems of equations now can be solved numerically, and the rest is similar to the analytical procedure presented above. Numerical results are shown in Fig. 7.1 and compared to the exact solution that was also found in the above.

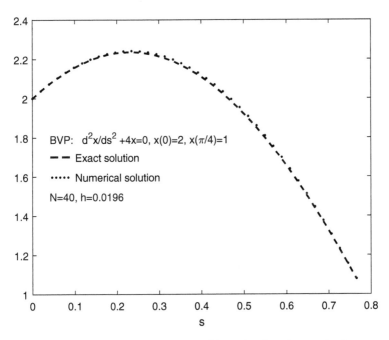

Fig. 7.1. Boundary value problem via shooting method.

MATLAB code for Fig. 7.1.

```
% inputting the initial information
N=50; a=0; b=pi/4;h= (b-a)/N;alpha=2; beta=1;
v(:,1)=[2 ;0]; u(:,1)=[0;1];
```

% finding the solution via shooting method
for n=1:N
t(n)= h*(n-1);v(:,n+1)=v(:,n)+h*[0 1; -4 0]*v(:,n);
u(:,n+1)=u(:,n)+h*[0 1; -4 0]*u(:,n);end
c=(beta-v(1,N+1))/u(1,N+1);
for n=1:N
x(n)=v(1,n)+c*u(1,n);xe(n)=2*cos (2*t(n))+sin(2*t(n)); end;
Ex=max(abs(x-xe));
plot(t,x,'k.', t,xe,'k–','linewidth',1.5)

In the above example, we presented a boundary value problem for a Homogeneous differential equation. Let's now present a BVP for following inhomogeneous differential equation:

$$Lx = \left(\frac{d^2}{ds^2} - 1 \right) x = 2e^s, \ s \in [0, 1], \quad \text{with}$$

$$x(0) = 1 \ \& \ x(1) = \frac{e^2 + 1}{e}$$

The exact solution to this problem is

$$x = se^s + e^{-s}$$

since $x' = e^s + se^s - e^{-s}$ and $x'' = 2e^s + se^s + e^{-s} = 2e^s + x$. The defined $x(s)$ satisfies $Lx = x'' - x = 2e^s$, and the boundary conditions, $x(0) = 1$ and $x(1) = \frac{e^2+1}{e}$. Thus, $x = se^s + e^{-s}$ is the solution.

Similar numerical procedure as used before is applicable, where y satisfying $y'' = y + 2e^s$ with $y(0) = 1$, $y'(0) = 0$, but z needs to satisfy $z'' = z$ with $z(0) = 0$ and $z'(0) = 1$.

Similarly define $x = y + cz$, where $c = \frac{e+e^{-1}-y(1)}{z(1)}$.

Next, write the above ODE in a vector form to find the numerical solutions

$$U = \begin{bmatrix} y \\ y' \end{bmatrix} = \begin{bmatrix} u_1 \\ u_2 \end{bmatrix} \Rightarrow U' = \begin{bmatrix} u_2 \\ u_1 + 2e^s \end{bmatrix} = \begin{bmatrix} 0 & 1 \\ 1 & 0 \end{bmatrix} U$$

$$+ \begin{bmatrix} 0 \\ 2e^s \end{bmatrix} = F(s, U)$$

and the second solution needs to satisfy

$$V = \begin{bmatrix} z \\ z' \end{bmatrix} = \begin{bmatrix} v_1 \\ v_2 \end{bmatrix} \Rightarrow V' = \begin{bmatrix} v_2 \\ v_1 \end{bmatrix} = \begin{bmatrix} 0 & 1 \\ 1 & 0 \end{bmatrix} V = \tilde{F}(V)$$

with $U(0) = \begin{bmatrix} 1 \\ 0 \end{bmatrix}$, and $V(0) = \begin{bmatrix} 0 \\ 1 \end{bmatrix}$.

Numerical results for this BVP using MATLAB is shown in Fig. 7.2.

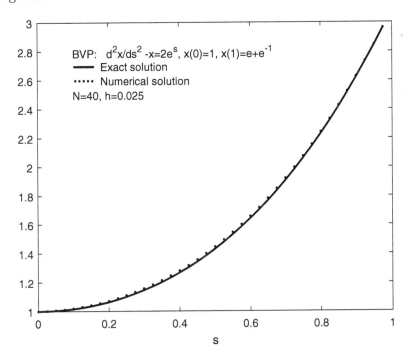

Fig. 7.2. BVP for inhomogeneous ODE.

MATLAB code for Fig. 7.2.

```
% inputting the initial information
N=40; a=0; b=1;h= (b-a)/N; alpha=1; beta=exp(1)+exp(-1);
u(:,1)=[alpha; 0]; v(:,1)=[0; 1];
% finding the solution via shooting method
for n=1:N
t(n)= h*(n-1); u(:,n+1)=u(:,n)+h*[0 1; 1 0]*u(:,n)+
```

[0;2*h*exp(t(n))];
v(:,n+1)=v(:,n)+h*[0 1; 1 0]*v(:,n); end;
c=(beta-u(1, N+1))/v(1,N+1);
for n=1:N
x(n)=u(1,n)+c*v(1,n); xe(n)=t(n)*exp(t(n))+1/exp(t(n)); end;
plot(t,x,'k.', t,xe,'k-','linewidth',1.3)

The same general procedure can be used to solve boundary value problems of the form

$$\frac{d^2x(s)}{ds^2} = p(s)\frac{dx(s)}{ds} + q(s)x(s) + r(s), \quad a \le s \le b$$

for other types of boundary conditions. Examples of such boundary values problems are provided in the exercise section at the end of this chapter.

It is apparent that error in shooting method solution is very much dependent on the method used to find numerical solutions for associated initial value problems. Let us as assume error in finding solutions to initial value problems are of order h^m. Applying error estimate procedure presented in Chapter 1, it follows the error in finding c will be $E_c = ||c - \tilde{c}||_\infty = O(h^m)$, where $c = \frac{\beta-y(b)}{z(b)}$ and $\tilde{c} = \frac{\beta-\tilde{y}(b)}{\tilde{z}(b)}$. The symbol 'tilde' over a variable indicates its numerically found value. Thus,

$$E_x = ||x - \tilde{x}||_\infty = ||y + cz - \tilde{y} - \tilde{c}\tilde{z}||_\infty = ||y - \tilde{y} + cz - \tilde{c}\tilde{z}||_\infty$$
$$= ||(y - \tilde{y}) + c(z - \tilde{z}) + (c - \tilde{c})\tilde{z}||_\infty = O(h^m)$$

In other words, the numerical accuracy of the solution to BVP found via the shooting method will be of the same order as the numerical accuracy of the procedure used to find IVP solutions numerically. For example, error $E_x = 0.0113$, associated Fig. 7.1, is in agreement with expectation that $E_x = O(h)$, since $h = 0.0198$.

7.2 Finite Difference Method for Boundary Value Problems

Another method to deal with BVP for linear second-order ODE, is the finite difference approximation. That is to approximate

$$x' = \frac{x(s_{k+1}) - x(s_{k-1})}{2h} + O(h^2)$$

$$x'' = \frac{x(s_{k+1}) - 2x(s_k) + x(s_{k-1})}{h^2} + O(h^2)$$

That is the differential equation $x'' = p(s)x' + q(s)x + r(s)$, is approximated by

$$\frac{x(s_{k+1}) - 2x(s_k) + x(s_{k-1})}{h^2}$$

$$= p(s_k)\frac{x(s_{k+1}) - x(s_{k-1})}{2h} + q(s_k)x_k + r(s_k) + O(h^2)$$

The above is a finite difference approximation of the desired ODE. Let's write it in its standard form

$$\frac{x_{k+1} - 2x_k + x_{k-1}}{h^2} = p_k\frac{x_{k+1} - x_{k-1}}{2h} + q_k x_k + r_k + O(h^2)$$

where $x_k = x(s_k)$ stands for the approximate value of exact solution $x(s_k)$ & $p_k = p(s_k)$, $q_k = q(s_k)$, $r_k = r(s_k)$, with $s_k = a + hn$, $n = 0, 1, 2, 3, \ldots, N$ and $h = \frac{b-a}{N}$. Factoring the common terms

$$\left(1 - \frac{h}{2}p_k\right)x_{k+1} - (2 + h^2 q_k)x_k + \left(1 + \frac{h}{2}p_k\right)x_{k-1} = h^2 r_k$$

Noting $x_0 = \alpha$ and $x_N = \beta$, and substituting them in above equations for $k = 1$ & $k = N - 1$, one finds

$$A\begin{bmatrix} x_1 \\ x_2 \\ \cdots \\ x_{N-2} \\ x_{N-1} \end{bmatrix} = \begin{bmatrix} -(1 + \frac{h}{2}p_1)\alpha + h^2 r_1 \\ h^2 r_2 \\ \cdots \\ h^2 r_{N-2} \\ -(1 - \frac{h}{2}p_{N-1})\beta + h^2 r_{N-1} \end{bmatrix} = \Gamma$$

where

$$A = \begin{bmatrix} a_{1,1} & a_{1,2} & 0 & \cdots & \cdots & 0 \\ a_{2,1} & a_{2,2} & a_{2,3} & \cdots & \cdots & 0 \\ 0 & a_{3,2} & a_{3,3} & a_{3,4} & \cdots & 0 \\ \vdots & \vdots & \vdots & \vdots & \vdots & \vdots \\ 0 & \cdots & \cdots & 0 & a_{N-1,N-2} & a_{N-1,N-1} \end{bmatrix}$$

$a_{k,k-1} = 1 + \frac{h}{2}p_k$, $a_{k,k} = -(2 + h^2 q_k)$ & $a_{k,k+1} = 1 - \frac{h}{2}p_k$.

In other words, A is a tridiagonal matrix with all its elements $a_{i,j}$ being zero except for $a_{i,i}$ and $a_{i,i\pm1}$. The unknowns $x_1, x_2, \ldots, x_{N-2}, x_{N-1}$, can be found by first finding A^{-1}. We have also seen if A is diagonally dominant

$$\left| -(2 + h^2 q_k) \right| > \left| 1 + \frac{h}{2}p_k \right| + \left| 1 - \frac{h}{2}p_k \right|$$

then the solution can be easily found by iteration even if N is large.

To see the performance of this method let us redo our previous example using finite difference procedure. That is consider the equation $x'' = -4x$, with $x(0) = 2$, $x(\frac{\pi}{4}) = 1$ and take $N = 40$ so that its results be a fair comparison of the previous result. Let us note for this example $p(s) = r(s) = 0$ and $q(s) = -4$. Hence, $a_{k,k\pm1} = 1$ and $a_{k,k} = -(2 - 4h^2)$, $\Gamma_1 = -\alpha$, $\Gamma_{N-1} = -\beta$ and $\Gamma_j = 0$, for $j = 2, 3, \ldots, N - 2$. Numerical results are shown in Fig. 7.3, which is finding same quantities as shown in Fig. 7.1, but using a different method.

MATLAB code for Fig. 7.3.

```
% inputting initial information
N=40; a=0; b=pi/4; h=(b-a)/N; alpha =2; beta=1;
N1=N-1;N2=N-2;
% finding exact solution
for n=1:N1
s(n)=a+h*n; ze(n)=2*cos (2*s(n))+sin(2*s(n)); end
% defining matrix A and finding solution via finite difference method
for i=1:N1
for jj=1:N1
A(i,jj)=0; end; end;
```

```
for i=1:N1
A(i,i)=-(2 − 4 ∗ h²);
ga(i)=0;
end;
ga(1)=-alpha; ga(N1)=-beta;
for i=1:N2
jj=i+1; A(i,jj)=1; end;
for i=2:N1
jj=i-1; A(i,jj)=1; end; z=inv(A)*ga';
plot(s,z,'k.', s,ze,'k−','linewidth',1.3 )
```

It has been shown that the error for finite difference approximation presented

$$E_x = ||x - \tilde{x}||_\infty = O(h^2)$$

See Keller (1976) for details of this analysis. We just note that the program providing Fig. 7.3 finds $E_x = 1.0717 * 10^{-04}$, which is consistent with estimate error $E_x = O(h^2)$, since $h^2 = 3.8553 * 10^{-04}$.

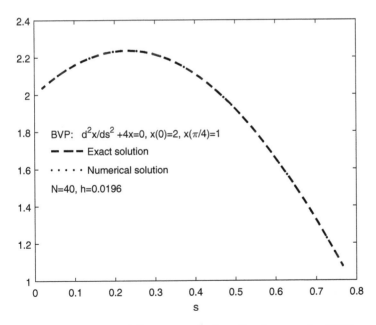

Fig. 7.3. Finite difference method application to solve BVP.

7.3 Shooting Method for Nonlinear Second-Order ODE Boundary Value Problems

The basic idea for solving nonlinear second-order ODE boundary value problem of the form:

$$x'' = f(s, x, x') \text{ with } x(a) = \alpha, x(b) = \beta \text{ for } a \leq s \leq b$$

is the same as what was presented for the linear case. To have a unique solution we need $f(s, x, x')$ to be a continuous function of s, x, x' and $\frac{\partial f}{\partial s}$, $\frac{\partial f}{\partial x}$ and $\frac{\partial f}{\partial x'}$ to be finite for all values of $s \in [a, b]$. The main difference between shooting method for linear equation and nonlinear equation is that the unknown $\gamma = x'(a)$ needed to be found from solution to the related IVP cannot be found as easily as we did in the linear case. One needs to find the appropriate γ via iteration. To see the procedure in more details, let us again start with finding a numerical solution to the initial value problem $x(s)$, with $x(a) = \alpha$ and $x'(a) = \gamma$, where γ is a constant chosen in such a way that $x(b)$ value is not too far from its desired boundary value $x(b) = \beta$. The formulated initial value solution depends on γ, and we may need to change γ in order to get an agreement between the found $x(b)$ and β. To emphasize this fact, let us denote the initial value solution of interest as $x(s, \gamma)$ and define:

$$g(\gamma) = x(b, \gamma) - \beta$$

and search for a γ that makes $g(\gamma) = 0$, or equivalently $x(b, \gamma) = \beta$. In other words, the procedure requires finding zero's of $g(\gamma)$. We have discussed several iterative procedures that can be used to find γ, but here we only describe how to apply Newton's iterative method to find γ, remembering our initial value problem is to satisfy:

$$x''(s, \gamma_j) = f(s, x(s, \gamma_j), x'(s, \gamma_j)) \quad \text{for } a \leq s \leq b$$

with $x(a, \gamma_j) = \alpha$ and $x'(a, \gamma_j) = \gamma_j$ for any choice of γ_j found via the Newton's procedure, which takes the form

$$\gamma_{j+1} = \gamma_j - \frac{g(\gamma_j)}{\dot{g}(\gamma_j)}, \quad j = 0, 1, 2, 3, \ldots$$

where $\dot{g}(\gamma) = \frac{dg(\gamma)}{d\gamma}$, and iteration is to stop when $|g(b, \gamma_j)| < \epsilon$, where ϵ is the desired tolerance level. Unlike standard application of Newton's method, $\dot{g}(\gamma)$ is not readily available and we need to solve another initial value problem to find $\dot{g}(\gamma)$. In order to find $\dot{g}(\gamma)$, take the derivative of $x''(s, \gamma)$ with respect to γ:

$$\frac{\partial x''(s, \gamma)}{\partial \gamma} = \frac{\partial f(s, x(s, \gamma), x'(s, \gamma))}{\partial \gamma} = f_x(s, x(s, \gamma), x'(s, \gamma)) \frac{\partial x(s, \gamma)}{\partial \gamma}$$

$$+ f_{x'}(s, x(s, \gamma), x'(s, \gamma)) \frac{\partial x'(s, \gamma)}{\partial \gamma},$$

In arriving at the above equation use was made of the fact that $\frac{\partial s}{\partial \gamma} = 0$. Defining $w(s, \gamma) = \frac{\partial x(s, \gamma)}{\partial \gamma}$ and assuming change of order of differentiation are justified for quantities under consideration, one finds

$$\frac{\partial x''(s, \gamma)}{\partial \gamma} = \frac{d^2}{ds^2} \frac{\partial x(s, \gamma)}{\partial \gamma} = \frac{d^2 w(s, \gamma)}{ds^2}$$

$$\equiv w''(s, \gamma) = f_x(s, x(s, \gamma), x'(s, \gamma)) w(s, \gamma)$$

$$+ f_{x'}(s, x(s, \gamma), x'(s, \gamma)) w'(s, \gamma)$$

where $w'(s, \gamma) = \frac{\partial x'(s, \gamma)}{\partial \gamma}$ and $w''(s, \gamma) = \frac{\partial x''(s, \gamma)}{\partial \gamma}$.

It also follows from the above that the differential equation of interest needs to satisfy the initial conditions

$$w(a, \gamma) = \frac{\partial x(a, \gamma)}{\partial \gamma} = \frac{\partial \alpha}{\partial \gamma} = 0 \quad \text{and}$$

$$w'(a, \gamma) = \frac{\partial x'(a, \gamma)}{\partial \gamma} = \frac{\partial \gamma}{\partial \gamma} = 1$$

Thus, finding $w(s, \gamma)$ reduces to finding the solution to the following second-order initial value problem numerically

$$w''(s, \gamma) = f_x(s, x(s, \gamma), x'(s, \gamma)) w(s, \gamma)$$

$$+ f_{x'}(s, x(s, \gamma), x'(s, \gamma)) w'(s, \gamma)$$

which satisfies $w(a, \gamma) = 0$ and $w'(a, \gamma) = 1$, as it was deduced in the above. Then evaluation of $w(s, \gamma)$ at $s = b$ allows us to determine

$$\dot{g}(\gamma) = \frac{dx(b, \gamma)}{d\gamma} = w(b, \gamma)$$

which is needed for applying the Newton's method numerically to find the zeros of $g(\gamma)$ need to find $x(s, \gamma)$ which makes $x(b, \gamma) = \beta$. In other words to find a γ_n that makes $g(\gamma_n) = 0$, one needs to apply the following Newton's iteration method

$$\gamma_{j+1} = \gamma_j - \frac{g(\gamma_j)}{\dot{g}(\gamma_j)} = \gamma_j - \frac{g(\gamma_j)}{w(b, \gamma_j)}, \quad j = 0, 1, 2, 3, \ldots$$

Of course, to start this iteration procedure, one needs to judiciously choose a starting value for γ_0 that results in a solution which is not too far away from true value of γ that leads to the desired boundary condition, just like standard Newton's method that the initial choice should not be too far from a zero of the function. To demonstrate this procedure let us consider the following boundary value problem:

$$x''(s) = -\frac{(x'(s))^2}{x(s)} + \frac{e^s}{x(s)}, \quad 0 \le s \le 1, x(0) = \sqrt{2},$$

$$\& \ x(4) = \sqrt{2e^4 + 8}$$

In this example,

$$f(s, x, x') = -\frac{(x')^2}{x} + \frac{e^s}{x}, \quad a = 0, \ b = 4, \ \alpha = x(a) = \sqrt{2},$$

$$\text{and } \beta = x(b) = \sqrt{2e^b + 2b}$$

It can be verified that exact solution to this problem is

$$x(s) = \sqrt{2e^s + 2s}$$

Let us now apply the above procedure to see how it performs for this example. The first initial problem we need to solve is

$$x'' = -\frac{(x')^2}{x} + \frac{e^s}{x}, \quad \text{with } x(0) = \sqrt{2} \ \& \ x'(0) = \gamma_0$$

where γ_0 is to be chosen judiciously. The second initial value problem needing to be solved is

$$w'' = f_x w + f_{x'} w' = \frac{(x')^2 - e^s}{x^2} w - \frac{2x'}{x} w', \text{ with}$$

$$w(0, \gamma) = 0 \ \& \ w'(0, \gamma) = 1$$

Using Euler's approximation we find solutions for $x(s, \gamma)$ & $w(s, \gamma)$ and apply Newton's method to find the desired γ, starting with initial value $\gamma_0 = 5$.

Figure 7.4 shows the result, which is a satisfactory solution to the nonlinear boundary value problem considered, and can be improved with higher N value or use of more accurate ODE solver.

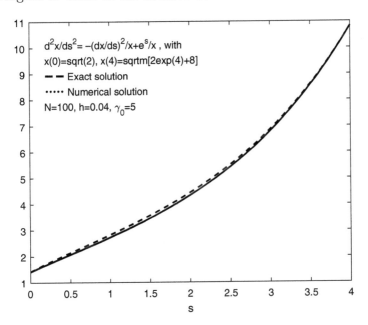

Fig. 7.4. BVP for nonlinear second-order ODE.

MATLAB code for Fig. 7.4.

```
% inputting initial information
N=100; a=0; b=4; h= (b-a)/N;alpha=sqrt(2);
beta=sqrt(2*exp(b)+2*b);
N1=N+1; M=10;ga(1)=5;q(:,1)=[0;1];
```

```
% computing the exact solution
for n=1:N1
t(n)=a+h*(n-1); xe(n)=sqrt(2*exp(t(n))+2*t(n)); end;
% finding x & w solutions via Euler's method
for k=1:M
x(:,1)=[alpha ;ga(k)];
for n=1:N
x(:,n+1)=x(:,n)+h*[x(2,n);(-(x(2,n))²+exp(t(n)))/x(1,n)];
w(:,n+1)= w(:,n)+h*[w(2,n); -(2*x(2,n)/x(1,n))*w(2,n)
+w(1,n)*((x(2,n))²-exp(t(n)))/(x(1,n))²]; end
% applying Newton's method
ga(k+1)=ga(k)- (x(1,N1)-beta)/w(1,N1); er(k)= x(1,N1)-beta; end;
for n=1:N1
z(n)=x(1,n); end
plot(t,z,'k-', t,xe,'k-','linewidth',1.3 )
```

For the example considered, sources of error include Euler's procedure to find the initial value solutions, and Newton's method to find suitable $\gamma = x'(a)$. Figure 7.4 indicates error in the found solution $E_x = ||x - \tilde{x}||_\infty = 0.1081$, which is of order $h = 0.0400$, the expected error $O(h)$, associated with the Euler's method used to generate (Fig. 7.4). For more details about boundary value error analysis, the reader is referred to Keller (1976).

Finite difference method can also be used to solve boundary value problems for nonlinear problems. The procedure is similar, except now one must find zeros of a system of nonlinear equations, by implementing a vector valued version of the Newton's method.

Exercises

7.1. (a) Find the exact solution to the following boundary value problem

$$x''+2x'-8x = 0 \text{ for } 0 \le t \le 2, \text{ with } x(0) = 2 \text{ and } x(2) = 0$$

(b) Use the shooting method along with Euler's procedure with $N = 20$, to find an approximate solution to the same boundary value problem.

(c) Compare your result to the exact solution.

7.2. (a) Find the exact solution to

$$(t^2 x')' - 2x = 0 \quad \text{for } t \in [1, e]$$

satisfying the boundary conditions $x(1) = 0$ & $x(e) = 1$, where $x' = \frac{dx}{dt}$.

Hint: Change variable t to $\tau = \ln t$ to transform the given ODE to a constant coefficient ODE.

(b) Apply the shooting method and Heun's procedure with $N = 20$, to find an approximate solution to the same boundary value problem.

(c) Compare your result to the exact solution.

7.3. Use the shooting method along with RK4 procedure with $N = 20$, to find an approximate solution to

$$x'' - \frac{x'}{t} - \frac{8x}{t^2} + 4 = 0, \text{ for } t \in [1, 2] \text{ and } x(1) = 1, \text{ & } x(2) = 10$$

Compare your result to the exact solution $x(t) = \frac{t^2(1+t^2)}{2}$.

7.4. (a) Find the exact solution to the following boundary value problem:

$$x'' + x = 0, \quad \text{for } 0 \le t \le 1, \quad \text{with } x'(0) = 1 \text{ & } x'(\pi/2) = 2$$

(b) Numerically find the approximate solution by using the shooting method along with Euler's procedure with $N = 20$.

(c) Compare your result to the exact solution you found in part (a).

Hint: Use the same idea as developed for the boundary conditions $x(a) = \alpha$ and $x(b) = \beta$.

7.5. Consider the boundary value problem

$$x'' - x = e^{2t}, \quad \text{for } 0 \le t \le 1, \; x(0) = 1 \text{ & } x'(1) = 0$$

(a) Find the solution using Heun's method with $N = 20$.

(b) Find the solution using RK2 method with $N = 20$.

(c) Compare your solutions in parts (a) and (b) and see if their errors are in the expected range.

7.6. Apply the shooting method to find approximate numerical solution to the boundary value problem

$$x'' - x = e^{2t}, \quad \text{for } 0 \le t \le 1$$

satisfying boundary conditions: $x(0) + x'(0) = 4e/3 + 1$ & $x(1) = e^2 + 1$.

Hint: Use Euler's method with $N = 20$ to find an approximate solutions to IVP: $x_1'' - x_1 = e^{2t}$ satisfying $x_1(0) = 1$ & $x_1'(0) = 0$ and an approximate solutions to IVP: $x_2'' - x_2 = 0$ satisfying $x_2(0) = 0$ & $x_2'(0) = 1$. Define $x = \alpha x_1 + \gamma x_2$ and apply the developed procedure for BVP.

7.7. Apply shooting method, using the Heun's method with $N = 20$, to find approximate solution to the BVP

$$\frac{d^2x(s)}{ds^2} + \frac{dx(s)}{ds} - 6x(s) = e^{-s}, \quad \text{for } 0 \le s \le 1$$

with $x'(0) = \frac{1}{6}$ and $x(1) + x'(1) = e^{-2}(\frac{9e^5}{4} - 1)$.

Hint: Use similar ideas as presented for BVP with boundary conditions $x(a) = \alpha$ and $x(b) = \beta$.

7.8. Same as Problem 7.1, except use finite difference method for the mentioned boundary value problem to find the approximate solution.

7.9. Same as Problem 7.2, except use finite difference method for boundary value problems to find the approximate solution.

7.10. Same as Problem 7.3, except use finite difference method for boundary value problems to find the approximate solution.

7.11. Find an approximate solution to the boundary value problem

$$x'' - x^2x' = 0 \quad \text{for } t \in [0, 3], \quad x(0) = \frac{1}{2} \,\&\, x(3) = \frac{1}{\sqrt{2}}$$

via the shooting method. Use Euler's procedure with $N = 20$.

7.12. Find an approximate solution to the boundary value problem

$$x'' - 2xx' + x'^2 = 0, \quad \text{for } t \in [0, 2]$$

with boundary conditions $x(0) = 2$ & $x(2) = 1 + e^4$, via the shooting method. Use Heun's procedure with $N = 20$.

7.13. Same as Problem 7.11, except use RK4 to find the numerical solution.

7.14. Find an approximate solution to the boundary value problem

$$x'' - 2xx' + x'^2 = 0, \text{ for } t \in [0, 2]$$

satisfying boundary conditions $x'(0) = 2$ & $x(2) = 1 + e^4$, via the shooting method. To find numerical solution, apply RK2 procedure with $N = 20$.

Hint: Use similar ideas as presented for BVP with boundary conditions $x(a) = \alpha$ and $x(b) = \beta$.

7.15. Same as Problem 7.13, except use Adams–Bashforth–Moulton with $N = 20$ to find its numerical solution. Use RK4 to start the Adams–Bashforth–Moulton procedure.

Chapter 8

Partial Differential Equations

Partial differential equations appear in many applied disciplines, since they are intimately related to what we deal with in everyday life, such as electricity, magnetism, heat, and waves. In this chapter, we mainly introduce partial differential equations of mathematical physics that have vast number of applications. The general form of second order partial differential equations of interest has the form

$$au_{xx} + bu_{xy} + cu_{yy} + du_x + eu_y + fu + g = 0$$

If $b^2 - 4ac < 0$ over the domain of interest, the equation is called *Elliptic*.

If $b^2 - 4ac > 0$ over the domain of interest, the equation is called *Hyperbolic*.

If $b^2 - 4ac = 0$ over the domain of interest, the equation is called *Parabolic*.

The partial differential equations of mathematical physics that we will consider have the following forms:

PDE type	Name	Equation
Elliptic	Poisson equation	$\frac{\partial^2 u}{\partial x^2} + \frac{\partial^2 u}{\partial y^2} = \rho(x, y)$
Hyperbolic	Wave equation	$\frac{\partial^2 u}{\partial t^2} - c^2 \frac{\partial^2 u}{\partial x^2} = 0$
Parabolic	Heat equation	$\frac{\partial u}{\partial t} - a^2 \frac{\partial^2 u}{\partial x^2} = 0$

Let us first introduce numerical procedures using finite difference method for solving each of the above PDE types.

8.1　Finite Difference Method for Elliptic Equations

As an example on how to numerically solve Elliptic equations using finite difference method, let us consider a function $u(x, y)$ satisfying the Poisson differential equation, with boundary condition given as $f(x, y)$. That is, $u_{xx} + u_{yy} = \rho(x, y)$ defined on rectangle $D = \{(x, y)|\ a < x < b\ \&\ c < y < d\}$, with boundary $S = \{(x, y)|\ x \in [a, b]$ for $y = c$ or $d\ \&\ y \in [c, d]$ for $x = a$ or $b\}$ and $u(x, y)$ on the boundary of D, satisfies the boundary condition

$$u(x, y) = f(x, y) \quad \text{for } (x, y) \in S$$

It is assumed that $\rho(x, y)$ and $f(x, y)$ are continuous over respective domains of their definition.

In order to apply finite difference method, one first partitions the interval $[a, b]$ into J equal parts of width $h = \frac{b-a}{J}$ and interval $[c, d]$ portioned into K equal parts of width $g = \frac{d-c}{K}$. Like what is done in one dimension, define $x_j = a + jh$ and $y_k = c + kg$ for $j = 0, 1, 2, \ldots, J$ and $k = 0, 1, 2, \ldots, K$, respectively. These defined points (x_j, y_k) are called mesh points.

Using central difference approximation for variable x one finds

$$\frac{\partial^2 u(x_j, y_k)}{\partial x^2} = \frac{u(x_{j+1}, y_k) - 2u(x_j, y_k) + u(x_{j-1}, y_k)}{h^2}$$

$$-\frac{h^2}{12} \frac{\partial^4 u(\xi_j, y_k)}{\partial x^4}$$

Likewise, for y

$$\frac{\partial^2 u(x_j, y_k)}{\partial y^2} = \frac{u(x_j, y_{k+1}) - 2u(x_j, y_k) + u(x_j, y_{k-1})}{g^2}$$

$$-\frac{g^2}{12} \frac{\partial^4 u(x_j, \gamma_k)}{\partial y^4}$$

where $\xi_j \in (x_{j-1}, x_{j+1})$ and $\gamma_k \in (y_{k-1}, y_{k+1})$. Combining these relations, discretized Poisson equation takes the following form:

$$u_{xx} + u_{yy} = \frac{u(x_{j+1}, y_k) - 2u(x_j, y_k) + u(x_{j-1}, y_k)}{h^2}$$

$$+ \frac{u(x_j, y_{k+1}) - 2u(x_j, y_k) + u(x_j, y_{k-1})}{g^2}$$

$$= \rho(x_j, y_k) - \frac{h^2}{12} \frac{\partial^4 u(\xi_j, y_k)}{\partial x^4} - \frac{g^2}{12} \frac{\partial^4 u(x_j, \gamma_k)}{\partial y^4}$$

Next, denoting $\rho_{j,k} = \rho(x_j, y_k)$, boundary condition $f(x, y)$ is written as $f_{j,k} = f(x_j, y_k)$, and label numerical approximate values of the solution as $u_{j,k} = u(x_j, y_j)$.

Using above notations, and application of the finite difference approximation to the Poisson equation leads to the following finite difference equation:

$$\beta_1 u_{j,k} - [u_{j+1,k} + u_{j-1,k}] - \beta_2 [u_{j,k+1} + u_{j,k-1}] = -\beta_3 \rho_{j,k}$$

where $\beta_1 = 2[1 + \frac{h^2}{g^2}]$, $\beta_2 = \frac{h^2}{g^2}$ & $\beta_3 = h^2$, with truncation error of order $O(h^2 + g^2)$.

Next, we require the approximate solution to satisfy the given boundary condition on S

$$u_{0,k} = f_{0,k} \quad \& \quad u_{J,k} = f_{J,k}, \quad k = 0, 1, 2 \ldots, K$$

and

$$u_{j,0} = f_{j,0} \quad \& \quad u_{j,K} = f_{j,K}, \quad j = 0, 1, 2 \ldots, J$$

Above information leaves us with $N = (J - 1)(K - 1)$ unknowns to find. For example when $J = 4$ and $K = 5$ there are 12 unknowns $u_{j,k}$, for $j = 1, 2, 3$ & $k = 1, 2, 3, 4$. They are found by writing out above found finite difference equation for values of j & k satisfying $1 \leq j \leq 3$ and $1 \leq k \leq 4$.

$$\beta_1 u_{j,k} - [u_{j+1,k} + u_{j-1,k}] - \beta_2 [u_{j,k+1} + u_{j,k-1}] = -\beta_3 \rho_{j,k}$$

Writing out the above equations for all allowed values of j & k, one finds

$$
\begin{bmatrix}
\beta_1 u_{1,1} - u_{2,1} - \beta_2 u_{1,2} \\
\beta_1 u_{1,2} - u_{2,2} - \beta_2[u_{1,3} + u_{1,1}] \\
\beta_1 u_{1,3} - u_{2,3} - \beta_2[u_{1,4} + u_{1,2}] \\
\beta_1 u_{1,4} - u_{2,4} - \beta_2 u_{1,3} \\
\beta_1 u_{2,1} - [u_{3,1} + u_{1,1}] - \beta_2 u_{2,2} \\
\beta_1 u_{2,2} - [u_{3,2} + u_{1,2}] - \beta_2[u_{2,3} + u_{2,1}] \\
\beta_1 u_{2,3} - [u_{3,3} + u_{1,3}] - \beta_2[u_{2,4} + u_{2,2}] \\
\beta_1 u_{2,4} - [u_{3,4} + u_{1,4}] - \beta_2 u_{2,3} \\
\beta_1 u_{3,1} - u_{2,1} - \beta_2 u_{3,2} \\
\beta_1 u_{3,2} - u_{2,2} - \beta_2[u_{3,3} + u_{3,1}] \\
\beta_1 u_{3,3} - u_{2,3} - \beta_2[u_{3,4} + u_{3,2}] \\
\beta_1 u_{3,4} - u_{2,4} - \beta_2 u_{3,3}
\end{bmatrix}
$$

$$
=
\begin{bmatrix}
u_{0,1} + \beta_2 u_{1,0} - \beta_3 \rho_{1,1} \\
u_{0,2} - \beta_3 \rho_{1,2} \\
u_{0,3} - \beta_3 \rho_{1,3} \\
u_{0,4} + \beta_2 u_{1,5} - \beta_3 \rho_{1,4} \\
\beta_2 u_{2,0} - \beta_3 \rho_{2,1} \\
-\beta_3 \rho_{2,2} \\
-\beta_3 \rho_{2,3} \\
\beta_2 u_{2,5} - \beta_3 \rho_{2,4} \\
u_{4,1} + \beta_2 u_{3,0} - \beta_3 \rho_{3,1} \\
u_{4,2} - \beta_3 \rho_{3,2} \\
u_{4,3} - \beta_3 \rho_{3,3} \\
u_{4,4} + \beta_2 u_{3,5} - \beta_3 \rho_{3,4}
\end{bmatrix}
= \vec{G}
$$

To find a matrix representation of the above that allows an easy way to find the unknowns $u_{j,k}$, one relabels the matrix $[u_{j,k}]$ as a column vector, where each row of matrix $[u_{j,k}]$ is written as a column vector and placed under each other successively. For our example with $J = 4$ & $K = 5$ the matrix $[u_{j,k}]$ is transformed to

$$
\vec{v} = [u_{1,1}, u_{1,2}, u_{1,3}, u_{1,4}, u_{2,1}, u_{2,2}, u_{2,3}, u_{2,4}, u_{3,1}, u_{3,2}, u_{3,3}, u_{3,4}]^\top
$$

where $u_{j,k}$ are relabeled as elements of the vector $\vec{v} = [v_1, v_2, \ldots, v_{11}, v_{12}]^\top$ with $v_{n(j,k)} = u_{j,k}$, and $n(j,k) = k + (-1 + j) * (K - 1)$. Such a relabeling of $u_{j,k}$ allows us to write the above system of equations as

$$A\vec{v} = \vec{G}$$

where

$$A = \begin{bmatrix}
\beta_1 & -\beta_2 & 0 & 0 & -1 & 0 & 0 & 0 & 0 & 0 & 0 & 0 \\
-\beta_2 & \beta_1 & -\beta_2 & 0 & 0 & -1 & 0 & 0 & 0 & 0 & 0 & 0 \\
0 & -\beta_2 & \beta_1 & -\beta_2 & 0 & 0 & -1 & 0 & 0 & 0 & 0 & 0 \\
0 & 0 & -\beta_2 & \beta_1 & 0 & 0 & 0 & -1 & 0 & 0 & 0 & 0 \\
-1 & 0 & 0 & 0 & \beta_1 & -\beta_2 & 0 & 0 & -1 & 0 & 0 & 0 \\
0 & -1 & 0 & 0 & -\beta_2 & \beta_1 & -\beta_2 & 0 & 0 & -1 & 0 & 0 \\
0 & 0 & -1 & 0 & 0 & -\beta_2 & \beta_1 & -\beta_2 & 0 & 0 & -1 & 0 \\
0 & 0 & 0 & -1 & 0 & 0 & -\beta_2 & \beta_1 & 0 & 0 & 0 & -1 \\
0 & 0 & 0 & 0 & -1 & 0 & 0 & 0 & \beta_1 & -\beta_2 & 0 & 0 \\
0 & 0 & 0 & 0 & 0 & -1 & 0 & 0 & -\beta_2 & -\beta_1 & -\beta_2 & 0 \\
0 & 0 & 0 & 0 & 0 & 0 & -1 & 0 & 0 & -\beta_2 & -\beta_1 & -\beta_2 \\
0 & 0 & 0 & 0 & 0 & 0 & 0 & -1 & 0 & 0 & -\beta_2 & -\beta_1
\end{bmatrix}$$

Above example shows the matrix A is a $12 \times 12 = N \times N$ square matrix made up of $4 \times 4 = (K-1) \times (K-1)$ blocks of following matrices:

$$I_{K-1} = \begin{bmatrix} 1 & 0 & 0 & 0 \\ 0 & 1 & 0 & 0 \\ 0 & 0 & 1 & 0 \\ 0 & 0 & 0 & 1 \end{bmatrix}, \quad L_{K-1} = \begin{bmatrix} 0 & 1 & 0 & 0 \\ 1 & 0 & 1 & 0 \\ 0 & 1 & 0 & 1 \\ 0 & 0 & 1 & 0 \end{bmatrix} \&$$

$$O_{K-1} = \begin{bmatrix} 0 & 0 & 0 & 0 \\ 0 & 0 & 0 & 0 \\ 0 & 0 & 0 & 0 \\ 0 & 0 & 0 & 0 \end{bmatrix}$$

that can be used to represent A as

$$A = \begin{bmatrix} S_{K-1} & -I_{K-1} & O_{K-1} \\ -I_{K-1} & S_{K-1} & -I_{K-1} \\ O_{K-1} & -I_{K-1} & S_{K-1} \end{bmatrix}$$

where $S_{K-1} = \beta_1 I_{K-1} - \beta_2 L_{K-1}$.

The same pattern holds for other values of $J\&K$ with the block submatrices being $(K-1) \times (K-1)$ square matrices, and matrix A made up of $(J-1)$ number of these blocks along its rows and columns, keeping the same pattern as shown above.

For small values of $J \& K$ Gaussian elimination can be used to find numerical solution $\vec{v}$ of the Poisson equation, but for large values of $J \& K$ iterative procedures need to be applied for finding the solution. To complete the example consider following PDE equation satisfying Poisson equation

$$\frac{\partial^2 u}{\partial x^2} + \frac{\partial^2 u}{\partial y^2} = 2x(2y^2 - 1)e^{-y^2}$$

defined over a rectangle with vertices at $(0,0), (0.8,0), (0.8,1)$ & $(0,1)$ and boundary condition

$$u(0,y) = 0, \ \ u(0.8, y) = 0.8e^{-y^2} \text{ for } 0 \le y \le 1$$

and

$$u(x,0) = x, \ \ u(x,1) = xe^{-1} \text{ for } 0 \le x \le 0.8$$

The exact solution to the above problem can be verified to be

$$u(x,y) = xe^{-y^2}$$

Table 8.1 shows numerical results for above example when $J = 4$ and $K = 5$. Note the expected error is of order

$$O(h^2 + g^2) = O(0.08)$$

Table 8.1 also shows the agreement between numerically found solution and the exact solution is within the accuracy of the procedure.

It should be noted that A^{-1} needs to be calculated in order to find the numerical solution, thus the eigenvalues of A should not be too close to zero in order not to have instability, which can magnify the error. For chosen values of $h \& g$ the eigenvalues of A all fall in the interval $[0.9678, 7.0322]$ making the inversion procedure to be stable. Table 8.1 shows the numerical solution and its difference from exact solution for the the above example.

Table 8.1. Solution for Poisson equation given in above example.

x	y	Numerical solution	Exact solution	Difference
0.2	0.2	0.1923	0.1922	−0.0002
0.2	0.4	0.1705	0.1704	−0.0001
0.2	0.6	0.1395	0.1395	0.0000
0.2	0.8	0.1054	0.1055	0.0001
0.4	0.2	0.3846	0.3843	−0.0003
0.4	0.4	0.3411	0.3409	−0.0002
0.4	0.6	0.2790	0.2791	0.0000
0.4	0.8	0.2108	0.2109	0.0002
0.6	0.2	0.5768	0.5765	−0.0003
0.6	0.4	0.5115	0.5113	−0.0002
0.6	0.6	0.4186	0.4186	0.0000
0.6	0.8	0.3162	0.3164	0.0002

MATLAB code for Poisson equation presented in Table 8.1 using finite difference method

```
% inputting initial information
J=4;  K=5;  J1=J-1;  K1=K-1;J2=J+1;  K2=K+1;  a=0;b=0.8;
c=0;d=1;
h=(b-a)/J; g=(d-c)/K; b1 = 2*(1+h²/g²); b2 = h²/g²; b3 = h²; N =
J1 * K1;
% specifying known boundary values
for j=1:J2
x(j)=a+(j-1)*h;u(j,1) = x(j)*exp(-c²); u(j,K2) = x(j)*exp(-d²);
end;
for j=1:J1
xi(j)=x(j+1); end;
for k=1:K2
y(k)=c+(k-1)*g; u(1,k) = x(1) * exp(-(y(k))²); u(J2,k) = x(J2) *
(-(y(k))²); end
% generating mesh points
for j=1:J2
for k=1:K2
x1(j,k)=a+(j-1)*h; y1(j,k)=c+(k-1)*g; end; end
```

```
% generating related finite difference matrices
for k=1:K1
yi(k)=y(k+1); end;
IK=[1 0 0 0; 0 1 0 0; 0 0 1 0; 0 0 0 1];
LK=[0 1 0 0; 1 0 1 0; 0 1 0 1; 0 0 1 0];
OK=[0 0 0 0; 0 0 0 0; 0 0 0 0; 0 0 0 0]; SK=b1*IK-b2*LK;
AK=[SK -IK OK;-IK SK -IK; OK -IK SK];
[vec, veig] = eig(AK);
% computing exact solution & the vector G=CK
for k=1:K2
for j=1:J2
u(j,k) = (x(j)) * exp(-(y(k))^2);
RO(j,k) = (4 * (y(k))^2 - 2) * (x(j)) * exp(-(y(k))^2); end; end;
CK=[u(1,2)+b2*u(2,1)-b3*RO(2,2); u(1,3)-b3*RO(2,3); u(1,4)-
b3*RO(2,4); u(1,5)+b2*u(2,6)-b3*RO(2,5); u(3,1)-b3*RO(3,2);
-b3*RO(3,3); -b3*RO(3,4); b2*u(3,6)-b3*RO(3,5); u(5,2)+b2*u(4,1)
-b3*RO(4,2); u(5,3)-b3*RO(4,3);
u(5,4)-b3*RO(4,4); u(5,5)+b2*u(4,6)-b3*RO(4,5)];
% finding the numerical solution
vv=inv(AK)*CK;
for k=1:K1
for j=1:J1
s(j,k)=k+(-1+j)*(K-1); vm2(j,k)=vv(s(j,k));
vm(j+1,k+1)=vm2(j,k); end;
end;
for k=1:K2
vm(1,k)=u(1,k); vm(J2,k)=u(J2,k); end
for j=1:J2
vm(j,1)=u(j,1); vm(j,K2)=u(j,K2); end;
for k=1:K1
for j=1:J1
p=j+1; r=k+1; uu(j,k)=u(p,r); uv(s(j,k))=uu(j,k); end; end;
for n=1:N
dif(n)=uv(n)-vv(n); nm(n)=n; end;
result=[ vv uv' dif']; uvm1=u-vm;
figure(1)
hold off; subplot (1,2,1)
mesh(x1,y1,vm,'linewidth',1.3);   xlabel('X-axis'),   ylabel('Y-axis'),
zlabel('Found solution')
shading flat; view(37, 20); lightangle (0, 0)
```

subplot (1,2,2)
mesh(x1,y1,uvm1,'linewidth',1.3)
xlabel('X-axis'), ylabel('Y-axis'), zlabel('Difference from exact')
shading flat; view(37, 40);lightangle (0, 0); hold on;

Figure 8.1 shows a three-dimensional graph of solution to the Poisson equation considered.

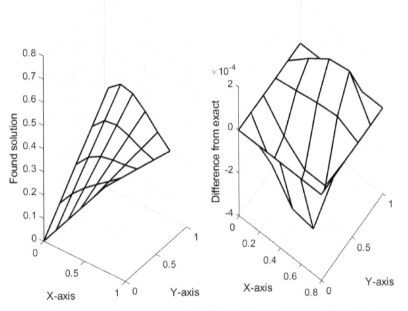

Fig. 8.1. Solution to Poisson equation via finite difference method. $u_{tt} + u_{xx} = 2x(2y^2-1)e^{-y^2}$, with $u(0,y) = 0, u(0.8,y) = 0.8e^{-y^2}, u(x,0) = x$ & $u(x,1) = xe^{-1}$.

8.2 Finite Difference Method for Hyperbolic Equations

The classical problem of the wave equation is the clamped string that considers how a disturbance propagates in the medium. In other words consider hyperbolic equation

$$\frac{\partial^2}{\partial t^2}u(x,t) - c^2 \frac{\partial^2}{\partial x^2}u(x,t) = 0, \quad \text{for } 0 < x < L, \& \ t > 0$$

with c being a constant in our presentation and $u(x,t)$ satisfying boundary conditions

$$u(0,t) = u(L,t) = 0, \quad \text{for } t > 0$$

and initial conditions

$$u(x,0) = \sigma_1(x) \;\&\; \frac{\partial u(x,0)}{\partial t} = \sigma_2(x), \quad \text{for } 0 \le x \le L$$

Similar to Elliptic equations one discretizes $x_j = jh$ and $t_k = kg$, $j = 1, 2, \ldots, J$ and $k = 1, 2, 3, \ldots$, where $h = \frac{L}{J}$ and g is increment of time unit of interest. Second derivatives also approximated the same as before to arrive at the following discretization:

$$u_{tt} - c^2 u_{xx} = \frac{u(x_j, t_{k+1}) - 2u(x_j, t_k) + u(x_j, t_{k-1})}{g^2}$$

$$- c^2 \frac{u(x_{j+1}, t_k) - 2u(x_j, t_k) + u(x_{j-1}, t_k)}{h^2}$$

$$= \frac{g^2}{12} \frac{\partial^4 u(x_j, \gamma_k)}{\partial t^4} - \frac{c^2 h^2}{12} \frac{\partial^4 u(\xi_j, t_k)}{\partial x^4}$$

Again denoting $u_{j,k}$ as numerical approximation of $u(x_j, t_k)$ we arrive at an approximate discretized version of the wave equation.

$$\frac{u_{j,k+1} - 2u_{j,k} + u_{j,k-1}}{g^2} - c^2 \frac{u_{j+1,k} - 2u_{j,k} + u_{j-1,k}}{h^2} = 0$$

with truncation error of order $O(h^2 + g^2)$. Simplifying the above one finds

$$u_{j,k+1} = -u_{j,k-1} + \gamma_1 u_{j,k} + \gamma_2 (u_{j+1,k} + u_{j-1,k})$$

where $\gamma_1 = 2(1 - \frac{c^2 g^2}{h^2})$ and $\gamma_2 = \frac{c^2 g^2}{h^2}$. To see how the initial conditions enter into the procedure, let us evaluate $u(j, 0) = \sigma_1(x_j)$ and apply Taylor theorem to find:

$$u(j, 1) = u(j, 0) + u_t(j, 0)g + \frac{1}{2} u_{tt}(j, 0)g^2 + O(g^3)$$

$$= \sigma_1(x_j) + \sigma_2(x_j)g + \frac{c^2 g^2}{2} \frac{d^2 \sigma_1(x)}{dx^2}\Big|_{x_j} + O(g^3)$$

In other words, initial conditions provides us with the information about $u_{j,0}$ & $u_{j,1}$ for $j = 0, 1, 2, \ldots, J$. This enables us to find

$$u_{j,2} = -u_{j,0} + \gamma_1 u_{j,1} + \gamma_2 (u_{j+1,1} + u_{j-1,1})$$

and other $u_{j,k}$'s for $j = 1, 2, \ldots, J$ & $k = 2, 3, \ldots, K$. Continuing the process developed, we are able to find numerically, the approximate values of $u(x_j, t_k)$ for all values of x_j & t_k of interest. To demonstrate the procedure in an actual example, let us consider the equation

$$u_{tt} - u_{xx} = 0$$

with boundary and initial conditions

$$u(\pi/2, t) = u(3\pi/2, t) = 0, \ u(x, 0) = 0 \ \& \ u_t(x, 0) = \cos x$$

It can be verified that that

$$u(x, t) = \cos x \sin t$$

is the exact solution to above problem. Knowing the exact solution, let us see how well the proposed method works for this problem. We take $J = K = 20$, and try to evaluate $u(x, t)$ at $t = 3$ in some given time unit. The selected choice of discretization leads to $h = (3\pi/2 - \pi/2)/J = 0.3142$ and $g = 3/K = 0.3$. According to approximation used, the expected error is of order $O(h^2 + g^2) = O(0.1887)$. Table 8.2 shows the numerical result for $u(x, t)$, for $t = 3$. It also shows numerical errors are within expectation.

Table 8.2. Solution for above wave equation at $t = 3$, using finite difference method.

x	Numerical solution	Exact solution	Difference
1.5708	0	0.0000	−0.0000
1.8850	−0.0446	−0.0436	−0.0010
2.1991	−0.0849	−0.0829	−0.0019
2.5133	−0.1168	−0.1142	−0.0027
2.8274	−0.1373	−0.1342	−0.0031
3.1416	−0.1444	−0.1411	−0.0033
3.4558	−0.1373	−0.1342	−0.0031
3.7699	−0.1168	−0.1142	−0.0027
4.0841	−0.0849	−0.0829	−0.0019
4.3982	−0.0446	−0.0436	−0.0010
4.7124	0	−0.0000	0.0000

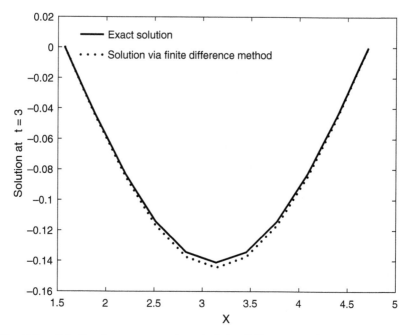

Fig. 8.2. Solution to wave equation via finite difference method at $t = 3$. $u_{tt} - u_{xx} = 0$, with $u(\pi/2, t) = u(3\pi/2, t) = 0$, $u(x, 0) = 0$ & $u_t(x, 0) = \cos x$.

MATLAB code for Table 8.2 and Fig. 8.2, using finite difference method

```
% inputting initial information
J=10; K=10;J2=J+1; K2=K+1;a=pi/2;b=3*pi/2;T=3; h=(b-a)/J;
g=T/K; c=1;
b1=2*(1-c² * g²/h²); b2=c² * g²/h²;
% inputting boundary and initial conditions
for j=1:J2
x(j)=a+(j-1)*h; u(j,1)=0; u(j,2)=u(j,1)+g*cos(x(j));end
for k=1:K2
t(k)=(k-1)*g;u(1,k)=0; u(J2,k)=0; end
% finding solution by iteration
for j=2:J
u(j,3)=-u(j,1)+b1*u(j,2)+b2*(u(j+1,2)+u(j-1,2)); end
for k=3: K
for j=2: J
```

u(j,k+1)=-u(j,k-1)+b1*u(j,k)+b2*(u(j+1,k)+u(j-1,k)) ;end;end
% finding exact solution and its difference
for j=1:J2
v(j)=u(j,K2); v1(j)=cos(x(j))*sin(t(K2)); dif(j)=v(j)-v1(j); end
Oerr=$g^2 + h^2$; res=[x' v' v1' dif']; plot(x,v,'k.',x,v1,'k')

8.3 Finite Difference Method for Parabolic Equations

The classical problem of the heat equation is to consider a rod of length L initially heated in a prescribed way, $f(x)$, and then one is to study it using finite difference method for parabolic equations to see how heat propagates. In other words consider parabolic equation with the following specified conditions:

$$\frac{\partial u}{\partial t} - \alpha^2 \frac{\partial^2 u}{\partial x^2} = 0, \ 0 < x < L \ \& \ t > 0$$

$$\text{with } u(x,0) = f(x) \ \text{ for } 0 \leq x \leq L$$

$$\text{and } u(0,t) = u(L,t) = 0 \ \text{ for } t > 0$$

The standard procedure is to again discretize x and t, $x_j = jh$ and $t_k = kg$ for $j = 0,1,2,\ldots,J$ & $k = 0,1,2,\ldots,K$ with $h = L/J$ and the choice for g will be discussed after finding the error estimate for this approximation is presented. However, we note for T to be the time when solution is to be found, then we need $K = \lceil T/g \rceil$. The standard choice of discretizing of $\frac{\partial^2 u}{\partial x^2}$ is like what was used in previous equations, but discretizing the $\frac{\partial u}{\partial t}$ leads to at least two choices. That is forward and backward difference approximations of the first derivative. It turns out if we use forward difference approximation, the procedure developed will not be stable, thus in this introductory presentation we use backward difference approximation, for the heat equation.

$$\frac{u(x_j, t_k) - u(x_j, t_{k-1})}{g}$$

$$-\alpha^2 \frac{u(x_{j+1}, t_k) - 2u(x_j, t_k) + u(x_{j-1}, t_k)}{h^2} = \tau_{j,k}$$

where the error

$$\tau_{j,k} = -\frac{g}{2}\frac{\partial^2}{\partial^2 t}u(x_j, \xi_k) + \alpha^2\frac{h^2}{12}\frac{\partial^4}{\partial^4}u(\xi_j, t_k) = O(g + h^2)$$

The above error estimate found indicates for the procedure to be sufficiently accurate it is best to choose $g = O(h^2)$. In addition, it has been shown that to have a stable numerical finite difference procedure, one needs to choose g in such a way that $\frac{\alpha^2 g}{h^2} \leq \frac{1}{2}$.

Like before, using the notation $u_{j,k}$ to approximately represent $u(x_j, t_k)$ one finds

$$u_{j,k} - u_{j,k-1} - \delta_1(u_{j+1,k} - 2u_{j,k} + u_{j-1,k}) = 0, \quad \text{where } \delta_1 = \frac{\alpha^2 g}{h^2}$$

Rearranging terms leads to following approximation of the Parabolic equation:

$$\delta_2 u_{j,k} - \delta_1(u_{j+1,k} + u_{j-1,k}) = u_{j,k-1}, \quad \text{where } \delta_2 = 1 + 2\delta_1$$

Similar to the wave equation procedure, we note that $u_{j,0} = f(x_j)$ is known for $j = 0, 1, 2, \ldots, J$.

Denote

$$\vec{v}_0 = [f_1, f_2, \ldots, f_{J-1}]^\top, \quad \text{where } f_j = f(x_j)$$

and define

$$\vec{v}_1 = [u_{1,1}, u_{2,1}, \ldots, u_{J-1,1}]^\top$$

Next, make use of the boundary conditions $u(0, k) = u(J, k) = 0$ to write the found discrete parabolic equation in a matrix form

$$A\vec{v}_1 = \vec{v}_0, \quad \text{where}$$

$$A = \begin{bmatrix} \delta_2 & -\delta_1 & 0 & 0 & 0 & 0 \\ -\delta_1 & \delta_2 & -\delta_1 & 0 & 0 & 0 \\ 0 & -\delta_1 & \delta_2 & -\delta_1 & 0 & 0 \\ \vdots & \vdots & \vdots & \vdots & \vdots & \vdots \\ 0 & 0 & 0 & -\delta_1 & \delta_2 & -\delta_1 \\ 0 & 0 & 0 & 0 & -\delta_1 & \delta_2 \end{bmatrix}$$

Noting that $J \times J$ matrix A is a diagonally dominated matrix, since $\delta_2 = 1 + 2\delta_1 \geq 2\delta_1$. It follows that even for large J values one can retrieve $\vec{v}_1$ from information on $\vec{v}_0$, iteratively. Let us write the solution as

$$\vec{v}_1 = A^{-1}\vec{v}_0$$

Having found $\vec{v}_1$ the procedure can be iteratively repeated to find

$$\vec{v}_k = A^{-1}\vec{v}_{k-1} = [A^{-1}]^k\vec{v}_0, \quad k = 1, 2, 3, \ldots$$

where $\vec{v}_k = [u_{1,k}, u_{2,k}, \ldots, u_{j,k}, \ldots, u_{J-1,k}]^\top$. It has been shown the eigenvalues of A are bounded below by 1, thus zero is not an eigenvalue, A^{-1} exists and presented procedure is stable. To give a numerical example let us first note one can rewrite

$$\frac{\partial u}{\partial t} - \alpha^2\frac{\partial^2 u}{\partial x^2} = 0 \quad \text{as} \quad \frac{\partial u}{\alpha^2\partial t} - \frac{\partial^2 u}{\partial x^2} = 0$$

Changing variable t to $\tau = \alpha^2 t$ above equation takes the form

$$\frac{\partial u}{\partial \tau} - \frac{\partial^2 u}{\partial x^2} = 0$$

which is equivalent to changing the time unit. Thus, for our next example we find numerical solution to above equation, with the following conditions:

$$u(x, 0) = \sin \pi x, \text{ for } 0 \le x \le 1$$

and

$$u(0, \tau) = u(1, \tau) = 0, \text{ for } \tau > 0$$

It can be verified that $u(x, \tau) = (\sin \pi x)e^{-\pi^2\tau}$ is the exact solution of this problem. To see numerical performance of the method, let us choose $J = 10$ and evaluate $u(x, \tau)$ at $\tau = 0.3, h = 1/J = 0.1$ and $g = h^2 = 0.01$. Numerical results for the above example is shown in Table 8.3. As it can be seen from Table 8.3, the difference between exact and found numerical solution is between 0.0011 and 0.0035, which is well within our expected error of $O(2h^2) \approx 0.02$, indicating a satisfactory performance of the proposed finite difference method.

Table 8.3. Solution for above parabolic equation at $\tau = 0.3$ using finite difference method.

x	Numerical solution	Exact solution	Difference
0.1000	0.0171	0.0160	0.0011
0.2000	0.0325	0.0304	0.0021
0.3000	0.0447	0.0419	0.0029
0.4000	0.0526	0.0492	0.0034
0.5000	0.0553	0.0518	0.0035
0.6000	0.0526	0.0492	0.0034
0.7000	0.0447	0.0419	0.0029
0.8000	0.0325	0.0304	0.0021
0.9000	0.0171	0.0160	0.0011

MATLAB code for Table 8.3 and Fig. 8.3, using finite difference method

```
% inputting initial information
J=10;J1=J-1; J2=J+1;JM=J1-1; a=0;b=1;T=1; h=(b-a)/J; g=h²;
T=0.3; K=ceil(T/g);
K2=K+1; d1=g/h²;d2=(1+2*d1);
% finding exact solution
for j=1:J1
x(j)=a+(j)*h; vk(j)=sin(pi*x(j)); end;
for k=1:K2
t(k)=(k-1)*g; end;
% Defining matrix A
for n=1: J1
for j=1: J1
A(j,n)=0; end; end;
for j=1:J1 A(j,j)=d2; end
for j=1:JM
A(j,j+1)=-d1; A(j+1,j)=-d1; end
for j=1:J1
v1(j)=sin(π*x(j))*exp(-π²*t(K2)); end wk=vk'; B=inv(A); C=B^K2;
wf = C*wk;
% finding numerical solution and its difference
dwv=wf-v1'; res=[x' wf v1' dwv]; plot(x,wf,'k.', x, v1,'k')
```

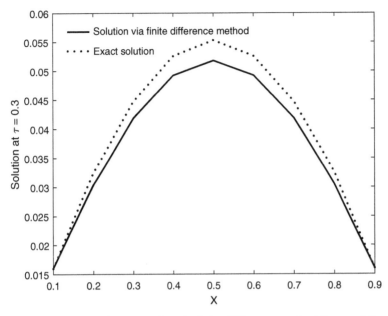

Fig. 8.3. Solution to heat equation via finite difference method for $\tau = 0.3$, $u_\tau - u_{xx} = 0$, with $u(x, 0) = \sin \pi x$, and $u(0, \tau) = u(1, \tau) = 0$.

Method of Lines (MoL)

Another procedure to find the solution of a PDE equation is the method of lines, where all but one of the independent variables are discretized. Such a procedure results in a system of differential equations, where numerical methods developed in Chapter 6 could be used to solve such a system of differential equations. To provide some concrete examples for solving PDE problems using method of lines let us apply the method to some of PDE problems considered in this chapter that used finite difference method for finding their solutions. We first start with the application of method of lines to parabolic equations since it results in a system of first order ODE without any need for modification.

8.4   Method of Lines for Parabolic Equations

Consider the heat equation, whose solution via matrix method was discussed in Section 8.3, but now we try to solve the problem using the method of lines.

$$\frac{\partial u}{\partial t} - \alpha^2 \frac{\partial^2 u}{\partial x^2} = 0, \ 0 < x < L \ \& \ t > 0$$

with

$$u(x,0) = f(x), \quad \text{for } 0 \le x \le L \ \text{and} \ u(0,t) = u(L,t) = 0$$

The easiest way to apply the method of lines is to discretize x and write it as $x_j = a + jh$, $j = 0, 1, \ldots, J$ & $h = L/J$. Then replace $\frac{\partial^2 u}{\partial x^2}$ by its standard finite difference approximation to find

$$\frac{du(x_j, t)}{dt}$$

$$-\alpha^2 \frac{u(x_{j+1}, t) - 2u(x_j, t) + u(x_{j-1}, t)}{h^2} = 0, \ j = 1, 2, \ldots, (J-1)$$

with $O(h^2)$ being order of truncation error for above approximation of parabolic equation, as discussed in Section 8.3.

Next, denote

$$\mathbf{U}(t) = [u(x_1, t), u(x_2, t), \ldots, u(x_j, t), \ldots, u(x_{J-1}, t)]^\top$$

and apply the boundary condition $u(x_0, t) = u(x_J, t) = 0$, to enable us write the above as the following system of ordinary differential equations:

$$\frac{d}{dt}\mathbf{U}(t) = A\mathbf{U}(t)$$

where

$$A = \gamma^2 \begin{bmatrix} -2 & 1 & 0 & \cdots & 0 & 0 & 0 \\ 1 & -2 & 1 & \cdots & 0 & 0 & 0 \\ 0 & 1 & -2 & 1 & \cdots & 0 & 0 \\ \vdots & \vdots & \vdots & \cdots & \vdots & \vdots & \vdots \\ 0 & 0 & 0 & \cdots & 1 & -2 & 1 \\ 0 & 0 & 0 & \cdots & 0 & 1 & -2 \end{bmatrix}, \quad \gamma = \frac{\alpha}{h}$$

Solution to above system of ODE can be verified to be

$$\mathbf{U}(t) = e^{At} * [f(x_1), f(x_2), \ldots, f(x_{J-1})]^\top$$

since

$$\frac{d}{dt}\mathbf{U}(t) = Ae^{At} * [f(x_1), f(x_2), \ldots, f(x_{J-1})]^\top = A\mathbf{U}(t)$$

and $\mathbf{U}(0) = [f(x_1), f(x_2), \ldots, f(x_{J-1})]^\top$, in agreement with the initial condition $u(x_j, 0) = f(x_j)$ for $j = 0, 1, 2, \ldots, J$. For this presentation $f(x_0) = f(x_J) = 0$ due to the selected boundary conditions. Thus, the above procedure provides numerical values of $u(x_j, t)$ for $j = 1, 2, \ldots, (J - 1)$, with the boundary conditions given as $u(x_0, t) = u(x_J, t) = 0$.

We should note that formal definition of an exponential with matrix exponent (denoted in MATLAB as expm) is given as

$$e^{At} = \sum_{n=0}^{\infty} \frac{(At)^n}{n!} = I + \frac{At}{1!} + \frac{A^2 t^2}{2!} + \frac{A^3 t^3}{3!} + \cdots$$

where I is the identity matrix. Thus,

$$\frac{d}{dt}e^{At} = A + \frac{2A^2 t}{2!} + +\frac{3A^3 t^2}{3!} + \cdots = A\left[I + \frac{At}{1!} + +\frac{A^2 t^2}{2!} + \cdots \right]$$

$$= A\sum_{n=0}^{\infty} \frac{(At)^n}{n!} = Ae^{At}$$

Indicating $\mathbf{U}(t)$ defined above is indeed the solution to the desired differential equation.

To compare performance of MoL to that using finite difference method, let us repeat the same numerical example we did using the finite difference method, but now use the MoL.

$$\frac{\partial u}{\partial \tau} - \frac{\partial^2 u}{\partial x^2} = 0$$

with

$$u(x, 0) = \sin(\pi x), \quad \text{and} \quad u(0, \tau) = u(1, \tau) = 0$$

Table 8.4 shows the numerical results, using MoL to find solution for above parabolic equation.

Table 8.4. Solution for above parabolic equation for $\tau = 0.3$, using MOL.

x	Numerical solution	Exact solution	Difference
0.1000	0.0164	0.0160	0.0004
0.2000	0.0312	0.0304	0.0007
0.3000	0.0429	0.0419	0.0010
0.4000	0.0504	0.0492	0.0012
0.5000	0.0530	0.0518	0.0013
0.6000	0.0504	0.0492	0.0012
0.7000	0.0429	0.0419	0.0010
0.8000	0.0312	0.0304	0.0007
0.9000	0.0164	0.0160	0.0004

MATLAB code for Table 8.4 and Fig. 8.4, using MoLs

```
% inputting initial information
J=10;J1=J-1; J2=J+1;JM=J1-1; a=0;b=1; h=(b-a)/J;g = h²;
T=0.3; K=ceil(T/g);K2=K+1; d1=1/h²;
% computing the exact solution
for j=1:J1
x(j)=a+(j)*h; f(j)=sin(pi*x(j)); end;
for k=1:K2
t(k)=(k-1)*g; end;
% Finding matrix A
for n=1: J1
for j=1: J1
A(j,n)=0; end; end;
for j=1:J1
A(j,j)=-2; end
for j=1:JM
A(j,j+1)=1; A(j+1,j)=1; end
for j=1:J1
v1(j)=sin(pi*x(j))*exp(-pi²*t(K2)); end ft=f'; B=d1*A;
% Finding solution via method of lines
ve=expm(B*t(K2))*ft; dv1=ve-v1'; res =[x' ve v1' dv1];
plot(x,v1,'k.', x, ve,'k')
```

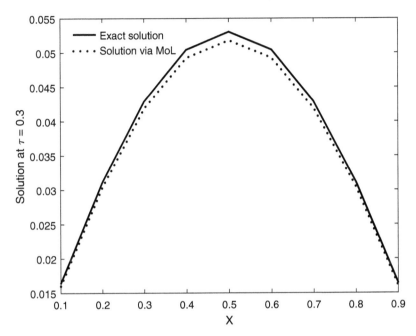

Fig. 8.4. Solution to heat equation via method of lines for $\tau = 0.3$, $u_\tau - u_{xx} = 0$, with $u(x,0) = \sin \pi x$, and $u(0,\tau) = u(1,\tau) = 0$.

Comparing Tables 8.3, and 8.4, Figs. 8.3 and 8.4, one notices that the numerical solution found using the method of lines is more accurate than the finite difference method. This is due to the fact that for the above example using method of lines, we did not need to approximate $\frac{\partial u}{\partial t}$ by a finite difference approximation in order to find the solution, but when using finite difference method, approximation of $\frac{\partial u}{\partial t}$ was needed. This finite difference approximation introduced additional truncation error in the found solution.

8.5 Method of Lines for Hyperbolic Equations

Let us again consider the wave equation for a clamped string

$$\frac{\partial^2}{\partial t^2} u(x,t) - c^2 \frac{\partial^2}{\partial x^2} u(x,t) = 0, \quad \text{for } 0 < x < L, \& t > 0, \text{ with}$$

$$u(0,t) = u(L,t) = 0 \text{ for } t > 0, \quad u(x,0) = \sigma_1(x) \ \& \ \frac{\partial u(x,0)}{\partial t} = \sigma_2(x)$$

To apply the MoL, let us discretize the variable $x_j = jh, j = 0, 1, 2, \ldots, J$ & $h = \frac{L}{J}$.

Next, let us apply the standard discretization to find

$$\frac{\partial^2}{\partial x^2} u(x, t) = \frac{u(x_{j+1}, t) - 2u(x_j, t) + u(x_{j-1}, t)}{h^2} + O(h^2)$$

Substitution of this approximation in the wave equation leads to

$$\frac{d^2}{dt^2} u(x_j, t) = \gamma^2 (u(x_{j+1}, t) - 2u(x_j, t) + u(x_{j-1}, t)),$$

$$j = 1, 2, \ldots, J - 1$$

where $\gamma = \frac{c}{h}$, and $O(h^2)$ is the order of its truncation error.

Writing the above in a matrix form with the vectorized solution denoted as

$$\mathbf{U}(t) = [u_1(t), u_2(t), u_j(t), \ldots, u_{J-1}](t)]^{\top}, \text{ where } u_j(t) = u(x_j, t)$$

we find

$$\frac{d^2 \mathbf{U}(t)}{dt^2} = A\mathbf{U}(t)$$

where

$$A = \gamma^2 \begin{bmatrix} -2 & 1 & 0 & \cdots & 0 & 0 & 0 \\ 1 & -2 & 1 & \cdots & 0 & 0 & 0 \\ 0 & 1 & -2 & 1 & \cdots & 0 & 0 \\ \vdots & \vdots & \vdots & \cdots & \vdots & \vdots & \vdots \\ 0 & 0 & 0 & \cdots & 1 & -2 & 1 \\ 0 & 0 & 0 & \cdots & 0 & 1 & -2 \end{bmatrix}$$

We now need to change this second-order ODE to a first-order ODE in order to apply numerical ODE methods that were developed.

$$\text{Defining} \quad \mathbf{w}(t) = \begin{bmatrix} \mathbf{U}(t) \\ \dfrac{d}{dt}\mathbf{U}(t) \end{bmatrix} = \begin{bmatrix} w_1(t) \\ w_2(t) \end{bmatrix}$$

the approximate form of the wave equation takes the form

$$\mathbf{w}'(t) = \frac{d}{dt} \begin{bmatrix} w_1(t) \\ w_2(t) \end{bmatrix} = \begin{bmatrix} w_2(t) \\ Aw_1(t) \end{bmatrix} = \begin{bmatrix} 0 & I \\ A & 0 \end{bmatrix} \mathbf{w}(t) = B\mathbf{w}(t)$$

where $B = \begin{bmatrix} 0 & I \\ A & 0 \end{bmatrix}$, with I being a $(J-1) \times (J-1)$ identity matrix, and 0 being a $(J-1) \times (J-1)$ zero matrix.

Next, let's note that $\mathbf{w}(0) = \begin{bmatrix} \mathbf{v_1} \\ \mathbf{v_2} \end{bmatrix}$

where $\mathbf{v_1} = [\sigma_1(x_1), \sigma_(x_1), \ldots, \sigma_1(x_{J-1})]^{\top}$ and $\mathbf{v_2} = [\sigma_2(x_1), \sigma_2(x_2), \ldots, \sigma_2(x_{J-1})]^{\top}$

with $\sigma_1(x_i)$ and $\sigma_2(x_i)$ are all given by the initial data. Similar to procedure used in Section 8.4, solution to the above ODE problem can be written as

$$\mathbf{w}(t) = e^{Bt}\mathbf{w}(0)$$

since matrix B is t independent.

To see the numerical performance of above procedure, let use consider the same example studied in Section 8.2.

$$u_{tt} - u_{xx} = 0$$

with $u(\pi/2, t) = u(3\pi/2, t) = 0$, $u(x, 0) = 0$ & $u_t(x, 0) = \cos x$, $J = 10$, and the solution to be evaluated at $t = 3$. The numerical result are shown in Table 8.5.

Table 8.5. Solution for above wave equation at $t = 3$ is found using the MoLs. The parameters are same as in Example 8.2, where finite difference method was used.

x	Numerical solution	Exact solution	Difference
1.8850	−0.0476	−0.0436	−0.0040
2.1991	−0.0905	−0.0829	−0.0075
2.5133	−0.1245	−0.1142	−0.0104
2.8274	−0.1464	−0.1342	−0.0122
3.1416	−0.1539	−0.1411	−0.0128
3.4558	−0.1464	−0.1342	−0.0122
3.7699	−0.1245	−0.1142	−0.0104
4.0841	−0.0905	−0.0829	−0.0075
4.3982	−0.0476	−0.0436	−0.0040

MATLAB code for Table 8.5 and Fig. 8.5 using MoLs

```
% inputting initial information
J=10; K=10;J1=J-1;a=pi/2;b=3*pi/2;T=3;JM=J1-1;
h=(b-a)/J;c=1; gama2=(c/h)^2;
% computing the exact solution
```

```
for j=1:J1
x(j)=a+j*h;  v1(j)=0;  v2(j)=cos(x(j));  esoT(j)=cos(x(j))*sin(T);
end; w0=[v1';v2'];
for j=1:J1
A1(j,j)=-2;
I(j,j)=1; end
for j=1:JM
A1(j,j+1)=1; A1(j+1,j)=1; end; A=gama2*A1;
for j=1:J1
for k=1:J1
z(j,k)=0; end; end;
% defining matrix B
B=[z I;A z];
% finding solution via method of lines
sol=(expm(B*T))*w0;
for j=1:J1
sol3(j)=sol(j); end; diff=sol3'-esoT'; res=[x' sol3' esoT' diff]
```

Truncation error for this example is $O(h^2) = 0.0987$, which is consistent with actual errors shown in Table 8.5. However, comparing numerical results found by applying finite difference method (Table 8.2 and Fig. 8.2) with the results found by the MoLs (Table 8.5 and Fig. 8.5) shows the for the example considered, finite difference method results are more accurate than the ones found using the method of lines. To make them have comparable error, one needs to use $J = 12$ subdivision for x when using the MoLs as compared to $J = 10$ when applying finite difference method to find the solution to above wave equation.

Although examples presented were simple applications of the MoLs, and solutions presented where mainly analytical, however numerical methods discussed in Chapter 6, could have been used to find associated numerical solution to the ODE system of equations, found via MoLs.

Examples presented also demonstrate the ability of MoL to reduce partial differential equations to systems of ordinary differential equations, that could make finding their solutions easier by applying procedures discussed for solving ordinary differential equations in previous chapters.

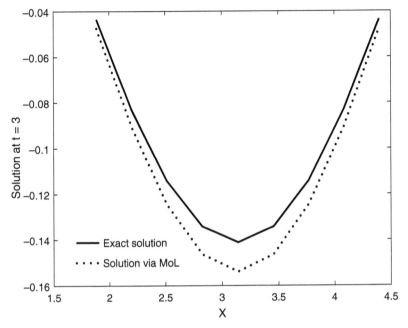

Fig. 8.5. Solution to wave equation via MoL at $t = 3$. $u_{tt} - u_{xx} = 0$, with $u(\pi/2, t) = u(3\pi/2, t) = 0$, $u(x, 0) = 0$ & $u_t(x, 0) = \cos x$.

Exercises

8.1. Use the finite difference method for elliptic equations to find an approximate numerical solution to the following partial differential equation:

$$u_{xx}(x, y) + u_{yy}(x, y)$$
$$= -\pi^2 \sin(x + y) - \pi^2 \sin(x - y) \text{ for } (x, y) \in S$$

where S is a square with vertices at $(0, 0), (1, 0), (1, 1)$ & $(0, 1)$, and boundary conditions $u(0, y) = u(1, y) = 0$ for $y \in [0, 1]$, $u(x, 0) = \sin(\pi x), x \in [0, 1]$ and $u(x, 1) = -\sin(\pi x)$, for $x \in [0, 1]$. Take $J = 5$, $K = 5$ and $h = \frac{1}{J}$.

Compare your approximate solution to the exact solution $u(x, y) = \sin(\pi x)\cos(\pi y)$ for mesh points $(x_j, y_k) = (jh, kh)$, $j = 0, 1, 2, \ldots, J$ & $j = 0, 1, 2, \ldots, K$.

8.2. Same a Problem 8.1, except solve for the Laplace equation.

$$u_{xx}(x,y) + u_{yy}(x,y) = 0$$

with boundary conditions $u(0,y) = -2y^2$, $u(1,y) = 2 - 2y^2$ for $y \in [0,1]$, $u(x,0) = 2x^2$ and $u(x,1) = 2x^2 - 2$ for $x \in [0,1]$.

Compare your result with the exact solution $u(x,y) = 2x^2 - 2y^2$ at the mesh points $(x_j, y_k) = (jh, kh)$, $j = 0, 1, 2, \ldots, J$ & $j = 0, 1, 2, \ldots, K$.

8.3. Use finite difference method for wave equations to find approximate solution $u(x, 2)$ of

$$\frac{\partial^2 u(x,t)}{\partial t^2} - \frac{1}{\pi^2} \frac{\partial^2 u(x,t)}{\partial x^2} = 0, \ t \geq 0, \ \& \ x \in [0,1]$$

satisfying $u(0,t) = 0$, $u(1,t) = 0$, $u(x,0) = \sin(\pi x)$ & $u_t(x,0) = 0$. Take $J = 15$, $K = 30$ and compare your approximate solution to the exact solution $u(x,t) = \sin(\pi x)\cos(t)$, at $t = 2$.

8.4. Same as Problem 8.3, except use the MoLs to find its solution.

8.5. Use finite difference method for wave equations to find the approximate solution of

$$\frac{\partial^2 u(x,t)}{\partial t^2} - \pi^2 \frac{\partial^2 u(x,t)}{\partial x^2} = 0, \ t \geq 0 \ \& \ x \in [0, \pi]$$

satisfying $u(0,t) = 0$, $u(\pi,t) = 0$, $u(x,0) = \sin x$ & $u_t(x,0) = \pi \sin(x)$. Take $J = 15, K = 5$ at $t = 1$.

Compare your approximate solution to the exact solution for this problem is $u(x,t) = \sin(x)(\sin(\pi t) + \cos(\pi t))$, at $t = 1$.

8.6. Same as Problem 8.5, except use the MoLs to find its solution.

8.7. Use finite difference method for wave equations to find the approximate numerical solution of

$$\frac{\partial^2 u(x,t)}{\partial t^2} - 4\frac{\partial^2 u(x,t)}{\partial x^2} = 0, \ \text{for} \ t \geq 0 \ \text{and} \ x \in \left[\frac{\pi}{2}, \frac{3\pi}{2}\right]$$

satisfying $u(x,0) = \cos(x)$, $u_t(x,0) = 0$, $u(\frac{\pi}{2}, t) = 0$ and $u(\frac{3\pi}{2}, t) = 0$.

Compare your approximate solution to the actual solution $u(x,t) = \cos(x)\cos(2t)$, for $J = K = 20$, at $t = 1$.

8.8. Same as Problem 8.7, except use the MoLs to find its solution.

8.9. Find the approximate solution $u(x,t)$ at $t = 0.2$ for

$$\frac{\partial u(x,t)}{\partial t} - \frac{\partial^2 u(x,t)}{\partial x^2} = 0, \ t \geq 0 \ \& \ x \in [1,3]$$

satisfying $u(1,t) = 0$, $u(3,t) = 0$ and $u(x,0) = \cos(\frac{\pi}{2}x)$. Use the finite difference method with $J = 15, K = 50$.

Compare your result with the exact solution $u(x,t) = \cos(\frac{\pi}{2}x)e^{-\frac{\pi^2}{4}t}$, at $t = 0.2$.

8.10. Same as Problem 8.9, except use the MoLs to find its solution.

8.11. Find the approximate solution $u(x,t)$ at $t = 0.2$ for

$$\frac{\partial u(x,t)}{\partial t} - \frac{1}{\pi^2}\frac{\partial^2 u(x,t)}{\partial x^2} = 0, \ t \geq 0 \ \& \ x \in [0,2]$$

satisfying $u(0,t) = 0$, $u(2,t) = 0$ & $u(x,0) = \cos(\frac{\pi}{2}(x+1)) + \cos(\pi(x+\frac{1}{2}))$. Use finite difference method with $J = 20, K = 20$.

Compare your result with the exact solution is $u(x,t) = e^{-t}\cos(\frac{\pi}{2}(x+1)) + e^{-4t}\cos(\pi(x+\frac{1}{2}))$, at $t = 0.2$.

8.12. Same as Problem 8.1, except use the MoLs to find its solution.

8.13. Develop the finite difference method for Parabolic equations to accommodate following conditions

$$\frac{\partial u}{\partial t} - \alpha^2 \frac{\partial^2 u}{\partial x^2} = 0, \ \ 0 < x < L, \text{ and } t > 0$$

satisfying $u(x,0) = \frac{x}{L}$ for $0 \leq x \leq L$, $u(0,t) = 0$, & u(L,t)=1, with $\alpha = 0.5$ and $L = 1$. Write a MATLAB code to accommodate these boundary/initial conditions. Take $J = 20, K=20$ for $t = 0.2$.

8.14. Modify the MoLs for parabolic equations to accommodate the case when α is not a constant, but a continuous function of x.

$$\frac{\partial u}{\partial t} - \alpha^2(x)\frac{\partial^2 u}{\partial x^2} = 0, \ 0 < x < 1 \ \& \ t > 0$$

with $u(x, 0) = \sin \pi x$ for $0 \le x \le 1$ and $u(0, t) = 0$, & $u(1, t) = 0$. Write a MATLAB code to accommodate the changed parabolic equation. Take $\alpha^2(x) = 1 + x$ and evaluate the solution $u(x, t)$ at $t = 0.5$.

8.15. Modify the MoLs for hyperbolic equations, to accommodate the case when c is not a constant but depends on x.

$$\frac{\partial^2}{\partial t^2}u(x, t) - c^2(x)\frac{\partial^2}{\partial x^2}u(x, t) = 0, \quad \text{for } 0 < x < 2, \& \ t > 0$$

and $u(0, t) = u(2, t) = 0$ for $t > 0$, $u(x, 0) = \sin x$ & $\frac{\partial u(x, 0)}{\partial t} = \cos x$, and $c^2(x) = 1 + e^{-x}$. Evaluate the solution $u(x, t)$ at $t = 1$.

References

Acton, F.S., *Numerical Methods that Work*, Harper and Row, 1970.

Ahlberg, J.H.E., Nilson, E.N., and Walsh, J.L., *The Theory of Splines and Their Applications*, Academic Press, 1967.

Ames, W.F., *Numerical Methods for Partial Differential Equations*, Academic Press, 1977.

Anton, H. and Rorres, C., *Elementary Linear Algebra with Applications*, John Wiley & Sons, 1989.

Atkinson, K.E., *An Introduction to Numerical Analysis*, John Wiley & Sons, 1991.

Baker, Jr. G.A. and Graves-Morris, P., *Padé Approximants*, Cambridge University Press, 1996.

Bellman, R., *Mathematical Methods in Medicine*, World Scientific, 1983.

Bellman, R. and Vasudevan, R., *Wave Propagation An Invariant Imbedding Approach*, D. Reidel Publishing Company, 1986.

Bleistein, N., *Mathematical Methods for Wave Phenomena*, Academic Press, 1984.

Blum, E.K., *Numerical Analysis and Computation: Theory and Practices*, Addison-Wesley, 1972.

Borse, G.J., *Numerical Methods with MATLAB, a Resource for Scientist and Engineers*, PWS, 1997.

Boyce, W.E. and DiPrima, R.C., *Elementary Differential Equations and Boundary Value Problems*, John Wiley & Sons, 2012.

Bradie, B., *A Friendly Introduction to Numerical Analysis*, Pearson Prentice-Hall, 2006.

Burden, R.L. and Faires, J.D., *Numerical Analysis*, PWS-Kent, 1989.

Chatelin, F., *Eigenvalues of Matrices*, Wiley, 1987.

Conte, S.D. and Boor, C.D., *Elementary Numerical Analysis, an Algorithmic Approach*, McGraw-Hill, 1972.

Davis, P. and Rabinowitz, P., *Method of Numerical Integration*, Academic Press, 1984.

Dennis, J.E. and Schnabel, R.B., *Numerical Methods for Unconstrained Optimization and Nonlinear Equations*, Prentice-Hall, 1983.

Dough, C.D., *Continued Fractions*, World Scientific, 2006.

Fausett, L.V., *Applied Numerical Analysis Using MATLAB*, Prentice-Hall, 1999.

Forsythe, G. and Wasow, W., *Finite Difference Methods for Partial Differential Equations*, Wiley, 1960.

Garabedian, P.R., *Partial Differential Equations*, John Wiley & Sons, 1964.

Gear, C.W., *Numerical Initial Value Problems in Ordinary Differential Equations*, Prentice Hall, 1971.

Gill, P., Murray, W., and Wright, M., *Practical Optimization*, Academic Press, 1981.

Golub, G. and Van Loan, C., *Matrix Computations*, Johns Hopkins Press, 1996.

Gregory, J. and Redmond, D., *Introduction to Numerical Analysis*, Jones and Bartlett, 1994.

Hageman, L. and Young, D., *Applied Iterative Methods*, Academic Press, 1981.

Hamermerlin, G. and Hoffman, K.H., *Numerical Mathematics*, Springer-Verlag, 1991.

Hanselman, D. and Littlefield, B., *Mastering MATLAB 7*, Pearson Prentice-Hall, 2005.

Herstein, L.N. and Winter, D.J., *A Primer on Linear Algebra*, Macmillan Publishing Company, 1988.

Hildebrand, F.B., *Introduction to Numerical Analysis*, McGraw-Hill, 1956.

Householder, A.S., *Principles of Numerical Analysis*, McGraw-Hill, 1953.

Hwon, Y.W. and Bang, H., *The Finite Element Method Using MATLAB*, CRC Press, 2000.

Isaacson, E. and Keller H., *Analysis of Numerical Methods*, John Wiley & Sons, 1966.

Johnson, C., *Numerical Solutions of Partial Differential Equations by Finite Element Method*, Cambridge University Press, 1987.

Keller, H., *Numerical Methods for Two-Point Boundary Value Problems*, Ginn (Blaisdell) Boston, 1968.

Kovach, L.D., *Boundary-Value Problems*, Addison-Wesley Publishing Company, 1984.

Kreyszig, E., *Advanced Engineering Mathematics*, John Wiley & Sons, 1999.

Kwon, Y.W. and Bang, H., *The Finite Element Method Using MATLAB*, CRC Press, 2000.

Lancaster, P. and Salkauskas, K., *Curve and Surface Fitting*, Academic Press, 1986.

Lanczos, C., *Applied Analysis*, Prentice-Hall, 1956.

Lapidus, L. and Seinfeld, J.H., *Numerical Solution of Ordinary Differential Equations*, Academic Press, 1971.

Linz, P. and Wang, R.L.C., *Exploring Numerical Methods: An Introduction to Computing Using MATLAB*, Jones and Bartlett, 2003.

Marsden, H.E. and Tromba, A.J., *Vector Calculus*, W.H. Freeman and Company, 1996.

Mathews, J.H., *Numerical Methods for Mathematics, Science, and Engineering*, Prentice-Hall, 1992.

Mathews, J.H. and Fink, K.D., *Numerical Methods Using MATLAB*, Prentice-Hall, 2004.

Mathews, J. and Walker, R.L., *Mathematical Methods of Physics*, W.A. Benjamin, 1964.

Miller, R.K. and Michel, A.N., *Ordinary Differential Equations*, Academic Press, 1982.

Milne, W.E., *Numerical Calculus*, Princeton University Press, 1949.

Nagle, R.K., Saff, E.B. and Snider, A.D., *Fundamentals of Differential Equations*, Addison-Wesley, 2000.

Ralston, A., *First Course in Numerical Analysis*, McGraw-Hill, 1965.

Razavy, M., *An Introduction to Inverse Problems in Physics*, World Scientific, 2020.

Recktenwald, G., *Numerical Methods with MATLAB*, Prentice-Hall, 2000.

Rice, J., *Numerical Methods, Software, and Analysis*, McGraw-Hill, 1983.

Richtmyer, R. and Morton, K., *Difference Methods for Initial Value Problems*, Wiley, 1967.

Schiesser, W.E. and Griffiths, G. W., *A Compendium of Partial Differential Equation Models: Method of Lines Analysis with Matlab*, Cambridge University Press, 2009.

Schumaker, L., *Spline Functions: Basic Theory*, Wiley, 1981.

Simmons, G.F., *Differential Equations with Applications and Historical Notes*, McGraw-Hill, 1991.

Strauss, W.A., *Partial Differential Equations An Introduction*, John Wiley & Sons, 1992.

Sundararajan, D., *The Discrete Fourier Transform: Theory, Algorithms and Applications*, World Scientific, 2001.

Weaver, H.J., *Theory of Discrete and Continuous Fourier Analysis*, John Wiley, 1989.

Widder, D., *The Heat Equation*, Academic Press, 1975.

Wouwer, A.V., Saucez, P., and Schiesser, W.E., *Adaptive Method of Lines*, Chapman & Hall/CRC, 2001.

Index

A

Adams–Bashforth method, 134
Adams–Bashforth–Multon method, 138
Adams–Moulton method, 136
adoptive quadrature methodology, 103

B

basic MATLAB commands, 1
Bolzano bisection method, 35
boundary value problems, 155
bracketing methods, 34

C

Cauchy–Peano existence theorem, 121
central difference approximation, 90
Chebyshev polynomials, 60
clamped spline, 65
closed quadrature, 95
composite quadrature, 99
composite Simpson quadrature, 101
composite trapezoidal quadrature, 100
condition number, 15
cubic splines, 64

D

diagonally dominated matrix, 19
discrete Fourier series, 83

E

Elliptic equations via finite difference method, 176
error-absolute, 4
error-relative, 4
Euclidean norm, 14
Euler's method, 119

F

fast Fourier transform, 82
Fehlberg method, 129
finite difference method for parabolic equations, 187
finite number of digits issues, 3
fixed point problem, 29
forward difference approximation, 90
Fourier series, 78

G

Gauss–Seidel method, 21
Gaussian elimination and back substitution, 11
Gibbs phenomenon, 80
global error, 124

H

Heun's method, 132
Horner's method, 6

I

Ill-conditioned matrix, 15
integration over infinite intervals, 109

J

Jacobi fixed point method, 46
Jacobi iteration method, 19
Jacobian, 45

L

Lagrange interpolating polynomials,
  57
least square, 69
linear convergence, 18
Lipschitz condition, 122
local error, 124
loss of significant digits, 4

M

matrix norms, 14
method of lines, 191
method of lines for hyperbolic
  equations, 195
method of lines for parabolic
  equations, 191
Milne–Simpson method, 140
modified Newton–Raphson methods,
  42
Monic Chebyshev polynomials, 60

N

natural spline, 65
Newton–Raphson method, 36
Newton–Raphson method for a
  system of equations, 50
numerical differentiation, 89
numerical stability, 140

O

open quadrature, 95
order symbol $O(h^n)$, 6

P

Parseval's identity, 82
partial pivoting, 11
precision degree, 94
principal value integrals, 107
propagation of errors, 5

Q

quadratic convergence, 18
quadrature of f, 94

R

rate of convergence, 17
rectangular quadrature, 96
Richardson extrapolation, 91
RK4 method, 129
root condition, 141
round-off errors, 4
Runge–Kutta method, 125

S

secant method, 39
shooting method, 157
Simpson rule, 99
singular/improper integrals,
  106
sparse matrix, 19
stability condition, 141
strongly stable procedures, 142

T

trapezoidal quadrature, 96
trigonometric polynomials, 80
truncation errors, 4

U

uniform norm, 14

V

vector norms, 14

W

wave equation, 183
weakly stable, 143

well-conditioned matrix,
16

Z

zeros of a function, 33